RESEARCH REPORT

Defense Planning Implications of Climate Change for U.S. Central Command

Karen M. Sudkamp, Elisa Yoshiara, Jeffrey Martini, Mohammad Ahmadi, Matthew Kubasak, Alexander Noyes, Alexandra Stark, Zohan Hasan Tariq, Ryan Haberman, Erik E. Mueller

NATIONAL DEFENSE RESEARCH INSTITUTE

For more information on this publication, visit **www.rand.org/t/RRA2338-5**.

About RAND

The RAND Corporation is a research organization that develops solutions to public policy challenges to help make communities throughout the world safer and more secure, healthier and more prosperous. RAND is nonprofit, nonpartisan, and committed to the public interest. To learn more about RAND, visit www.rand.org.

Research Integrity

Our mission to help improve policy and decisionmaking through research and analysis is enabled through our core values of quality and objectivity and our unwavering commitment to the highest level of integrity and ethical behavior. To help ensure our research and analysis are rigorous, objective, and nonpartisan, we subject our research publications to a robust and exacting quality-assurance process; avoid both the appearance and reality of financial and other conflicts of interest through staff training, project screening, and a policy of mandatory disclosure; and pursue transparency in our research engagements through our commitment to the open publication of our research findings and recommendations, disclosure of the source of funding of published research, and policies to ensure intellectual independence. For more information, visit www.rand.org/about/principles.

RAND's publications do not necessarily reflect the opinions of its research clients and sponsors.

Published by the RAND Corporation, Santa Monica, Calif.

Library of Congress Cataloging-in-Publication Data is available for this publication.
ISBN: 978-1-9774-1248-5

ABOUT THIS REPORT

THIS REPORT EXAMINES how U.S. Central Command (CENTCOM) planners could use operations, activities, and investments (OAIs) in the coming decades to address security threats that are related to stressors from climate change. We begin by considering OAIs that could help prevent—or mitigate the intensity of—conflict related to climate hazards. Understanding that, even with preventive action, the command will face additional requirements from climate stress, we then analyze the frequency and conditions under which the United States has traditionally intervened militarily globally and present rough order of magnitude costs of interventions by type (e.g., stabilization, humanitarian assistance and disaster relief). The aim of this report is to help CENTCOM planners prepare for a future security environment that is affected by climate change.

This report is the fifth in a series stemming from a larger project considering the impacts of climate change on the security environment in the region. The first report, *A Hotter and Drier Future Ahead: An Assessment of Climate Change in U.S. Central Command*, presents an analysis of projected climate impacts in the CENTCOM area of responsibility (AOR) in 2035, 2050, and 2070. The second report, *Pathways from Climate Change to Conflict in U.S. Central Command*, details causal pathways from climate change to conflict, including cases in which those pathways have played out in the CENTCOM AOR. *Conflict Projections in U.S. Central Command: Incorporating Climate Change*, the third report in the series, presents a range of forecasts of future conflict in the region, and climate change is incorporated as one driver of that conflict. The fourth report, *Mischief, Malevolence, or Indifference? How Competitors and Adversaries Could Exploit Climate-Related Conflict in the U.S. Central Command Area of Responsibility*, presents an analysis of how competitors—China, Russia, and Iran—might attempt to exploit climate-related conflict in the region.

The research reported here was completed in July 2023 and underwent security review with the sponsor and the Defense Office of Prepublication and Security Review before public release.

RAND National Security Research Division

This research was sponsored by CENTCOM and conducted within the International Security and Defense Policy Program of the RAND National Security Research Division (NSRD), which operates the National Defense Research Institute (NDRI), a federally funded research and development center sponsored by the Office of the Secretary of Defense, the Joint Staff, the Unified Combatant Commands, the Navy, the Marine Corps, the defense agencies, and the defense intelligence enterprise.

For more information on the RAND International Security and Defense Policy Program, see www.rand.org/nsrd/isdp or contact the director (contact information is provided on the webpage).

Acknowledgments

We thank Brian Michael Jenkins and Bruce Nardulli of the RAND Corporation for their thorough review of this report. Their review and constructive feedback improved the quality of the analysis. We additionally recognize Auburn Brown, Flannery Dolan, Krista Romita Grocholski, Alan Pino, and Howard Shatz for the research and analysis they conducted in support of this report. Consultations with Nathan Chandler were critical for the execution of our rough order of magnitude costing analysis. We also extend our gratitude to Jessica Arana and Chandra Garber, who brought their design and research communications expertise to make this report more accessible to readers; we also thank Kristen Meadows for her assistance on the design of this report. Finally, we thank Rosa Maria Torres, who assisted with document formatting.

CONTENTS

FIGURES AND TABLES

Figures

Tables

REPORT 1
Climate Hazards and Impacts

REPORT 2
Conflict Pathways

REPORT 3
Conflict Projections

REPORT 4
Adversarial Responses

5

REPORT 5
Defense Planning Implications

- Identify off-ramps to conflict and requirements
- Identify likely intervention types
- Analyze likely costs associated with interventions

KEY FINDINGS

OVER THE COMING DECADES, stressors from climate change will become more intense and more frequent in the U.S. Central Command (CENTCOM) area of responsibility (AOR). This development will likely contribute to CENTCOM's broader shift from a warfighting-focused command to a command that responds to and conducts both traditional and nontraditional security missions.

- The causal pathways from climate hazards to conflict revolve around political and economic concerns. Therefore, CENTCOM will likely play a supporting role to interagency partners in reducing the risk of climate-related conflict. However, military-led operations, activities, and investments provide some niche tools to interrupt progression along those pathways and could decrease the severity of conflicts by improving U.S. and partner response capabilities.
- In addition to mitigating conflict risk, CENTCOM has an opportunity to develop partner resilience to climate hazards with the ancillary benefit of strengthening bonds within the CENTCOM coalition. The United States has an interest in broadening cooperation between established CENTCOM partners and its newest coalition member, Israel, which was designated as part of CENTCOM in 2021.
- In the coming decades, there is likely to be an increased demand on CENTCOM to directly support humanitarian assistance and disaster response (HADR) in the theater and help partners build their own response capabilities.
- If climate stress leads to more conflicts in the CENTCOM AOR and U.S. policymakers define U.S. interests in ways that lead Washington to intervene, then stabilization operations would likely drive the highest costs for the operation types considered in the costing analysis.
- HADR, counterterrorism operations, noncombatant evacuation operations, and planned security cooperation (i.e., from the base budget rather than overseas contingency operations funding) impose lesser costs on Department of Defense funding.

CHAPTER 1

INTRODUCTION

THIS REPORT ADDRESSES how U.S. Central Command (CENTCOM) planners can use operations, activities, and investments (OAIs) to prevent—or mitigate the intensity of—climate-related conflict. We recommend that CENTCOM leverage regional partnerships and innovation opportunities related to climate resilience and adaptation to support nontraditional security cooperation activities in the region. This is the final report in a series of five. In our first report, *A Hotter and Drier Future Ahead: An Assessment of Climate Change in U.S. Central Command*, we identified extreme heat, water availability, and extreme precipitation as the primary climate hazards regional countries will face over the coming half century. Many of these hazards lead to compounding impacts. As the area of responsibility (AOR) experiences more extreme heat days with decreasing amounts of precipitation, there will be an increased demand for water that further strains freshwater supplies for human consumption. A resulting drought can exacerbate extreme heat because the land has a reduced ability to aid in cooling. Or, as the soil becomes increasingly arid, there is a greater risk of water runoff that leads to flash flooding when precipitation does occur. Limited governance around resource management in a resource-constrained region could also exacerbate local and regional tensions as populations and economies feel more negative effects from climate change.[1]

Our second and third reports addressed how climate could contribute to future conflict in the AOR. The analysis found that climate hazards can act as a threat multiplier, interacting with other drivers of conflict to increase the incidence or intensity of conflict. In the second report, *Pathways from Climate Change to Conflict in U.S. Central Command*, we identified causal pathways that often begin with climate hazards and result in a form of insecurity (e.g., food, livelihood, or physical or health insecurity) that combines with impacts on state capacity, population flows, and other factors. These impacts, when filtered through individual and armed group incentives to mobilize around greed or grievance, can culminate in conflict.[2]

Building on that research, we then used a machine learning framework to generate conflict projections for the AOR out to 2070. The third report, *Conflict Projections in U.S. Central Command: Incorporating Climate Change*, focused on the frequency of conflict in the theater. We found that, under a variety of socioeconomic and climate conditions, the CENTCOM AOR is projected to experience substantial conflict in the coming half century. Although there is suggestive evidence that worse climate outcomes will correlate with a greater incidence of conflict between 2040 and 2060, temperature increases and declines in precipitation are not the major drivers of the security environment according to our modeling, which indicated that governance concerns, prior conflict experience, and economic underdevelopment will remain the primary drivers of conflict. That said, there are good reasons to believe that existing research and our own conflict forecasts might be underestimating the impact of climate variables on conflict because of the model's reliance on historical patterns that might not occur in the future.[3]

To help address how shifting climates could affect the future security environment, we asked experts to consider how China and Russia—the principal international competitors of the United States—and primary U.S. regional adversary, Iran, could attempt to exploit climate change to advance their own security interests in the theater. Report four, *Mischief, Malevolence, or Indifference? How Competitors and Adversaries Could Exploit Climate-Induced Conflict in the U.S. Central Command Area of Responsibility*, presented scenarios that the project team developed of future climate-related crises. A group of subject-matter experts then analyzed adversary and competitor reactions to these scenarios during a two-day workshop. The workshop participants did not believe that China, Russia, or Iran would necessarily respond to climate-related conflict any differently than they would to conflict that was unrelated to climate, but there were climate-related variables those countries could exploit. Finally, the workshop highlighted that conflicts could occur along shared geographic combatant command boundaries, complicating the U.S. response and potentially forcing Washington to choose which partner to support in militarized interstate disputes.[4] Figure 1.1 illustrates the report series, and Figure 1.2 depicts the CENTCOM AOR.

Building on the prior research in this series, this final report addresses the defense planning implications of climate change for CENTCOM. Climate change, along with other transnational threats, are often discussed as part of a broader concept known as *nontraditional security*.[5] Many of the threats that are part of the nontraditional security concept, such as infectious disease and large-scale migration, are exacerbated by climate change. This study examined which traditional military tools can be applied to this nontraditional security threat and which new tools, which we call *nontraditional security cooperation*, can be developed to address the implications of climate change for CENTCOM.

Chapter 2 examines how CENTCOM can support regional partners to build resilience to climate stress. The chapter begins by identifying OAIs that could operate as off-ramps along the causal pathways from climate hazards to conflict by averting the conflict, reducing its intensity, or providing more-favorable circumstances for intervention. In addition to the off-ramp analysis, the chapter presents the results of a structured brainstorm in which subject-matter experts (area specialists and experts in service-specific military capabilities) generated a list of OAIs CENTCOM could undertake that are aligned with three overarching requirements driven by climate stress in the theater. We determined that these requirements incorporate climate impacts into

1. U.S. partner posture and planning
2. CENTCOM partner posture and planning
3. preparation for an increase in demand for humanitarian assistance and disaster response (HADR).

Finally, Chapter 2 discusses why addressing climate change is an opportunity to advance U.S. security interests in the region and is not just a threat that imposes requirements on CENTCOM as the relevant geographic combatant command.

The third chapter provides a discussion of historic patterns and indicators of the use of military force by the United States.

Figure 1.1. Progression of Reports in This Series

REPORT 1
Climate Hazards and Impacts
- Identify climate hazards
- Conduct climate analysis

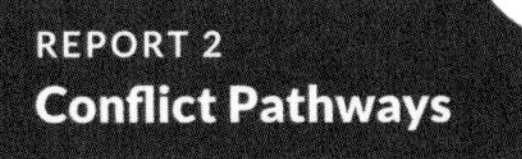

REPORT 2
Conflict Pathways
- Identify causal pathways to conflict
- Identify human system impacts and insecurities
- Determine types of conflict

REPORT 3
Conflict Projections
- Determine interplay between climate and socioeconomic projections and conflict data

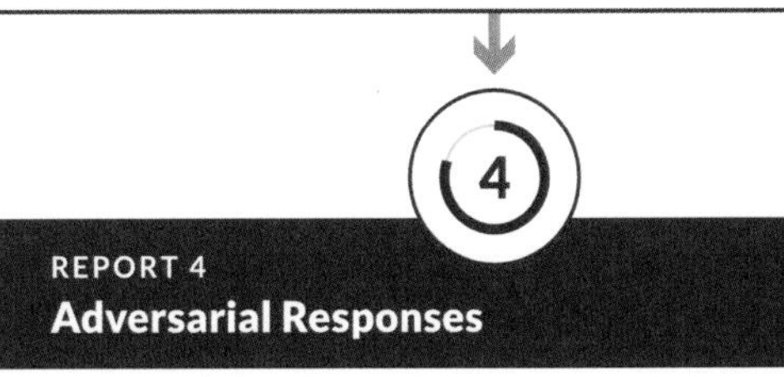

REPORT 4
Adversarial Responses
- Identify the regional interests of China, Russia, and Iran
- Identify responses to climate-influenced conflict
- Identify non-military responses to climate hazards

REPORT 5
Defense Planning Implications
- Identify off-ramps to conflict and requirements
- Identify likely intervention types
- Analyze likely costs associated with interventions

Figure 1.2. The U.S. Central Command Area of Responsibility

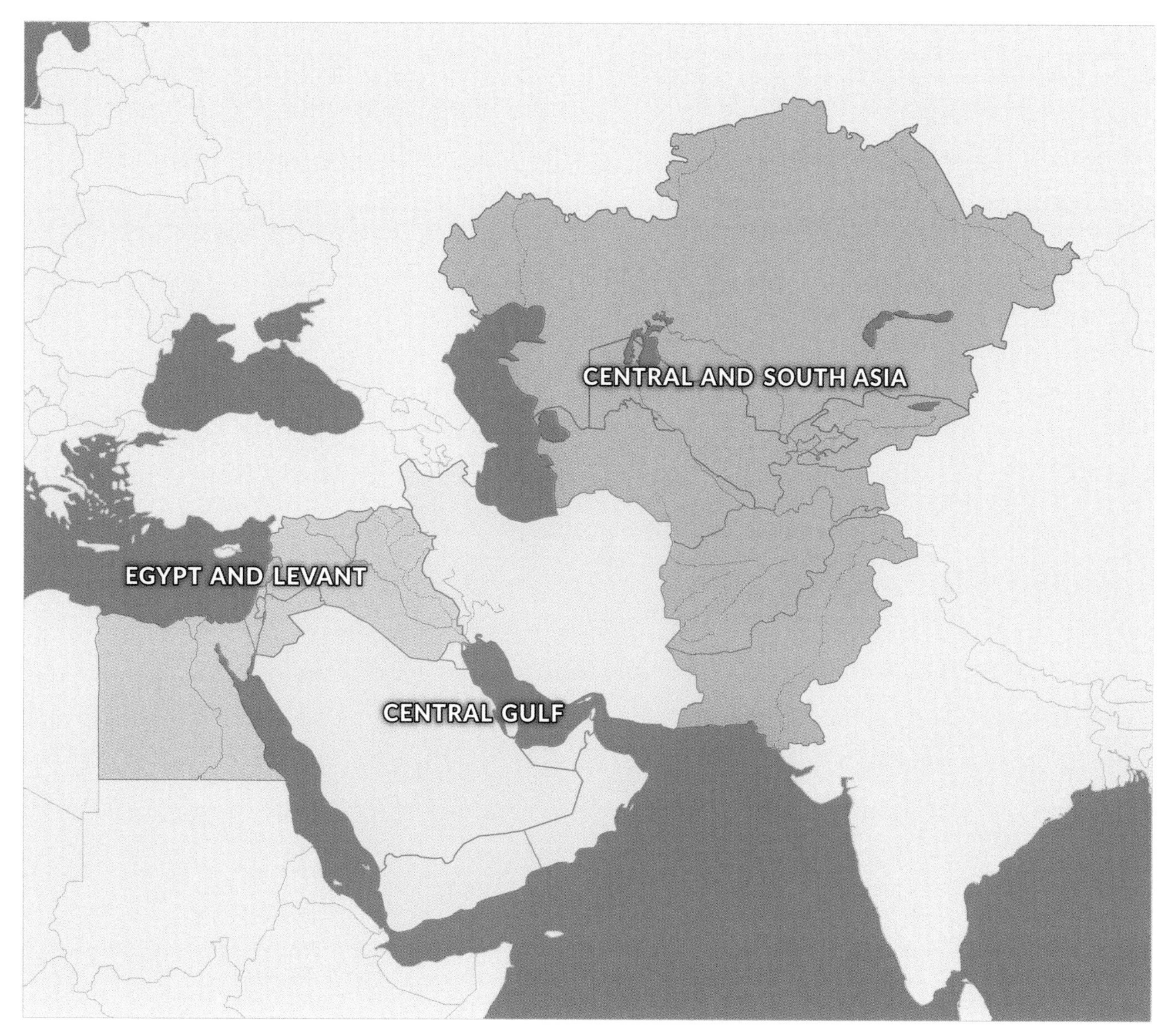

SOURCE: RAND-designed graphic based on email communications with CENTCOM, October 12, 2022.

The purpose of this analysis is to help CENTCOM anticipate potential demands for different operation types (e.g., stabilization, HADR) that our research suggests could increase as a result of climate change. Because defense planning is obliged to match requirements to the availability of forces and resources, we couple this analysis with rough order of magnitude (ROM) cost estimates for these operations. This analysis should inform CENTCOM planners as they consider how best to meet future requirements in a climate-affected security environment.

Endnotes

1 Michelle E. Miro, Flannery Dolan, Karen M. Sudkamp, Jeffrey Martini, Karishma V. Patel, and Carlos Calvo Hernandez, *A Hotter and Drier Future Ahead: An Assessment of Climate Change in U.S. Central Command*, RAND Corporation, RR-A2338-1, 2023.

2 Nathan Chandler, Jeffrey Martini, Karen M. Sudkamp, Maggie Habib, Benjamin J. Sacks, and Zohan Hasan Tariq, *Pathways from Climate Change to Conflict in U.S. Central Command*, RAND Corporation, RR-A2338-2, 2023.

3 Mark Toukan, Stephen Watts, Emily Allendorf, Jeffrey Martini, Karen M. Sudkamp, Nathan Chandler, and Maggie Habib, *Conflict Projections in U.S. Central Command: Incorporating Climate Change*, RAND Corporation, RR-A2338-3, 2023.

4 Howard J. Shatz, Karen M. Sudkamp, Jeffrey Martini, Mohammad Ahmadi, Derek Grossman, and Kotryna Jukneviciute, *Mischief, Malevolence, or Indifference? How Competitors and Adversaries Could Exploit Climate-Related Conflict in the U.S. Central Command Area of Responsibility*, RAND Corporation, RR-A2338-4, 2023.

5 NTS-Asia, "About Non-Traditional Security," webpage, undated.

CHAPTER 2

BUILDING PARTNER RESILIENCE TO CLIMATE-RELATED CONFLICT

CLIMATE HAZARDS AND STRESSORS will create a variety of challenges for U.S. partners by exacerbating water and food insecurity, contributing to migration within and from the region, and increasing pressure on state capacity to respond to climate-related disasters. The United States has an interest in building partner resilience to the full array of these challenges, but Washington's foremost security interest is preventing these challenges from resulting in armed conflict. This chapter begins by identifying off-ramps from our second report, which documents causal pathways from climate hazards to conflict.

We then complement that analysis with a second effort to generate a broader array of OAIs that would increase the preparedness of CENTCOM and its partners for climate challenges. That output is based on a structured brainstorm with nine RAND Corporation experts with experience in joint military operations and interventions, security cooperation, and regional expertise. The experts identified OAIs aligned with three overarching defense requirements driven by climate stress in the AOR that CENTCOM could consider fielding. Those requirements incorporate climate impacts into

1. U.S force posture and planning
2. CENTCOM partner force posture and planning
3. preparation for increased demand for HADR operations.

CENTCOM defines OAIs as follows:

> United States Central Command (USCENTCOM) directs and enables military operations and activities with allies and partners to increase regional security and stability in support of enduring U.S. interests. USCENTCOM executes a military campaign plan across the Central Region AOR through the application of operations, activities, and investments (OAIs) in support of steady state, crisis, and contingency operations. USCENTCOM activities include bilateral and multilateral exercises, combined interoperability training, professional education, senior military interaction, security cooperation, humanitarian assistance, and other traditional military activities. USCENTCOM activities mandate building strong relationships with regional leaders and maintaining access to shared facilities within the AOR.[1]

The third and final component of this chapter broadens the aperture even further, considering how adaptation to climate change could be leveraged by CENTCOM for an ancillary benefit: to promote cooperation among states within the AOR. A key aspect of deepening this type of cooperation is reinforcing Israel's position as a new, as of 2021, member of the coalition. The initiatives described in this section focus on sharing technology and scientific information, an area in which Israel is one of the most advanced CENTCOM partners.

Operations, Activities, and Investments as Conflict Off-Ramps

The second report in this series, *Pathways from Climate Change to Conflict in U.S. Central Command*, presented an analysis of the causal pathways from climate hazards to potential conflict and differentiated between two potential outcomes: intrastate conflict and interstate conflict.[2] Figure 2.1 depicts the last three steps in the *intrastate* conflict causal pathway, starting with what we termed the *critical pathway juncture* that comes after climate hazards lead to various forms of insecurity (e.g., food, health, physical insecurity). Our distillation of existing literature found that these insecurities combine with other pressures (e.g., diminished state capacity, migration) at critical pathway junctures that can propel a state along the path to internal conflict. It is also at this step in the causal pathway that military OAIs—whether undertaken by the United States or a foreign partner—can help stave off the conflict progression.

As they pertain to reducing the risk or impact of climate-related conflict, military-led OAIs can achieve several objectives. The first objective is to halt the path toward armed conflict, thus averting a potential conflict from occurring. The second is to mitigate the factors that could drive more lethal or longer duration conflicts, thus reducing the intensity of the conflict that ensues. And the third objective is to create the conditions for a more successful intervention if local partners or the United States elect to intervene in the resulting conflict. Of course, it is also possible that neither the local partner nor the United States will elect to intervene militarily, judging that it is not within their interests or capacity to do so.

The Department of Defense (DoD) is not the most logical U.S. government (USG) agency to address the first critical pathway juncture: state fiscal or political crisis, decline in legitimacy and capacity. Building fair and responsive governance, inoculating a state to fiscal shocks, and building capacity at all levels of government are the obvious ways to address this critical pathway juncture, which largely rest outside DoD's mandate. That said, DoD does have a niche but still important role to play in building local capacity to respond to climate pressures. One of the causes of declining state legitimacy in this context is poor government response to climate-related natural disasters, such as floods. For example, an OAI in which CENTCOM worked through the National Guard State Partnership Program (SPP) to develop counterpart capabilities to effectively respond to more frequent and more-intense natural disasters is one way to lower the risk of this critical pathway juncture progressing to armed conflict. Among the geographic combatant commands (GCCs)—exempting U.S. Northern

Figure 2.1. Final Steps in Causal Pathways to Intrastate Conflict

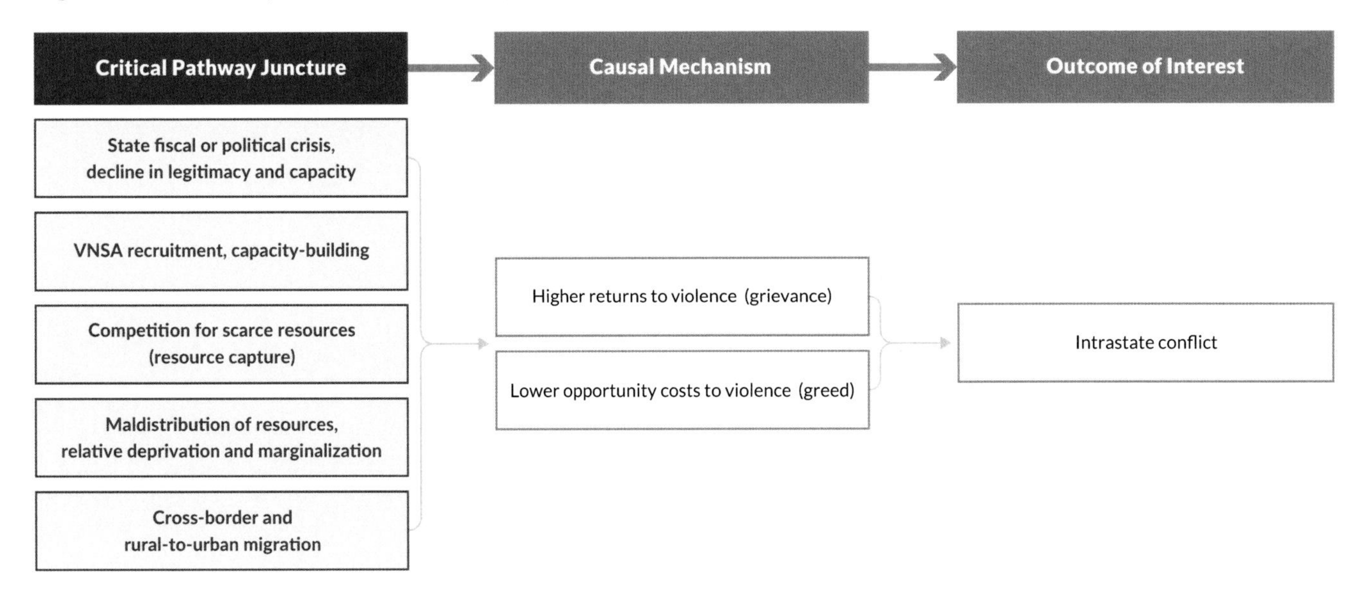

SOURCE: This figure was adapted from Chandler et al., 2023.

NOTE: VNSA = violent nonstate actor.

Command (NORTHCOM), which is not eligible for the SPP—CENTCOM has the lowest number of SPPs, with just nine such partnerships.[3] And several of the most disaster-prone countries in the CENTCOM AOR, such as Pakistan (which is vulnerable to tropical cyclones and monsoons) are among the countries with no SPP. Pairing these CENTCOM partners with a state National Guard that has similar disaster response experience is one off-ramp that could interrupt the causal pathway progression from climate hazard to armed conflict.

A second potential OAI that would reduce the risk of climate-related conflict via the same critical pathway juncture is investment in more resilient logistics (e.g., redundancy in road networks) that could otherwise cripple disaster response by creating choke points. Our research suggests that extreme precipitation is likely to increase the most in Tajikistan, Oman, and Yemen compared with the historical baseline. Combined with aridity, extreme precipitation will increase the risk of flooding in these areas. U.S. Army Central is already developing concepts to increase the resilience of logistics in the so-called Western Access Network, with many of that network's nodes clustering in western Saudi Arabia (e.g., Yanbu, Tabuk, Taif) and Oman (Salalah, Duqm). This creates an opportunity to develop—contingent on host country support—dual-use infrastructure (i.e., civilian and military applications) that provides in-depth defense against the increased range, precision, and abundance of Iranian firepower, while serving a second goal of enhancing disaster response in areas that will become increasingly at risk for extreme weather events. For example, developing Yanbu as a seaport of disembarkation could help make force flow more resilient to the Iranian threat and provide facilities that are well suited to disaster response along the Red Sea coast of Saudi Arabia.

For the second critical pathway juncture—an increase in VNSA recruitment and capacity-building—a potential mitigating OAI could be foreign partners organizing information operations that reveal the true motivations behind the VNSAs' championing of peoples who are facing climate-related hardships, such as economic dislocation, food insecurity, or forced migration. Special Operations Command Central has robust capabilities in psychological operations in the theater that could be employed to expose VNSA attempts to exploit climate-related disasters for their recruitment and fundraising. U.S. advising could also be employed to reinforce the partner's communication of its disaster response efforts, bolstering partner government legitimacy when warranted. Finally, CENTCOM could consider expanding civil affairs engagements and activities to assist foreign partners in building resilient government capacity in communities that are facing climate-related hazards, thus reducing the appeal of VNSAs as an alternative to the host country government.

Because both the third and fourth critical pathway junctures are about the distribution of resources among groups, improvements in partner security governance are a logical way to reduce the risk of conflict progression. Although security forces have limited influence over the size of the resource pie that groups compete for, their role in interdicting contraband, distributing relief, and conveying equal concern for the security of all societal groups can curtail negative forms of resource competition. Militaries are increasingly being called on to respond to climate disasters and to provide additional training for foreign partners to increase their capacity.[4] Finally, improved border security, including the capacity to provide for displaced populations inside a partner state, can mitigate the risk that migration feeds conflict. These example OAIs are visualized in Figure 2.2 as off-ramps on the causal pathway to intrastate conflict.

Although the research process of developing off-ramps from the causal pathways identified in the existing literature proved promising for intrastate conflict, we found the approach less promising for interstate conflict; therefore, we forgo a similar exercise for that conflict type here. There are two main reasons why this exercise is less conducive to interstate conflict. The first is that the conflict type and its most intense manifestation—interstate war—remain very rare phenomena relative to intrastate conflict.[5] This is not to imply that interstate war does not have significant consequences when it does occur—the largest numbers of battle deaths in the region coincided with the devastating 1980–1988 Iran-Iraq War—but these rare events create fewer opportunities to establish clear patterns in conflict progression, which is reflected in the substantially thinner treatment of this conflict type in our literature review that generated the causal pathways. The second reason is that what we synthesized as critical pathway junctures from that literature do not easily translate into military-led OAIs that would serve as off-ramps. For example, interstate competition over freshwater resources is perhaps the most often discussed pathway to interstate war, but the logical off-ramp for stopping that progression is the negotiation of water-sharing agreements between riparian countries, which is not a military mission.

Figure 2.2. Intrastate Conflict Causal Pathway Off-Ramps

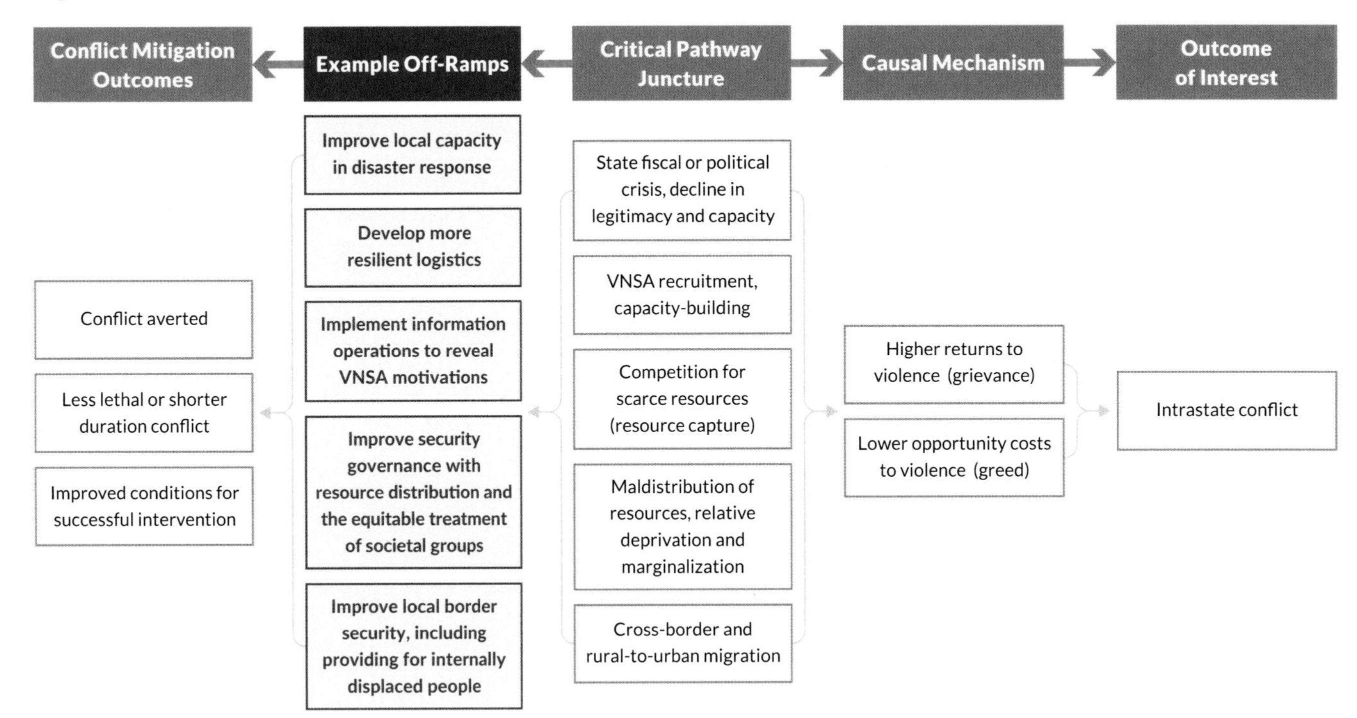

Additional Operations, Activities, and Investments Supporting Preparedness for Climate Change

Through a structured brainstorming session with nine RAND subject-matter experts, we identified additional OAIs in Table 2.1 that could increase regional cooperation and security, mitigate the risk of climate-related instability, and contribute to climate change preparedness. We have organized these OAIs to support three overarching requirements that incorporate climate impacts into (1) U.S. and (2) CENTCOM partner posture and planning and (3) preparing for an increase in demand for HADR. We also provided implementation time frame estimates for the OAIs. Short term is in the next one to three years, medium term is three to seven years, and long term is seven to ten years. The purpose of this exercise was to support CENTCOM in its OAI development process as it considers climate-informed actions in the AOR to mitigate the risk of climate-related conflict and improve the preparedness and resiliency of regional partners and allies.

Climate Change Provides Regional Partnership and Innovation Opportunities

Up to this point, we have framed CENTCOM's role as managing the security threats that emanate from climate hazards. However, climate change is not just a threat for the command; it also presents an opportunity to leverage the scientific and technological strengths of the United States and those of CENTCOM's regional partners, such as Israel and the United Arab Emirates (UAE), to deepen the CENTCOM coalition by building partner resilience to climate stress. This logic is consistent with Line of Effort 5 from the *Department of Defense Climate Adaptation Plan*, which emphasizes how responses to climate change can be an opportunity for DoD and such GCCs as CENTCOM. Line of Effort 5 calls on DoD to "enhance adaptation and resilience through collaboration" and, in addition to collaboration with U.S. interagency partners, highlights relationships with foreign partners as a critical area for cooperation:

> Build partner nation capacity to respond to climate change-related hazards. Actively participate in technical, academic, and scientific exchanges to

Table 2.1. Security Cooperation-Related Operations, Activities, and Investments

Requirement	OAIs	Time Frame
Incorporate climate impacts into U.S. force posture and planning in the AOR	Incorporate climate hazards into regional exercises, including Bright Star, the Juniper series, and the International Maritime Exercise, using JTEEP funds when appropriate.	Short term
	Request that DoD align high-demand capabilities for climate-related contingencies to CENTCOM. This could include civil affairs units for stabilization operations or intratheater lift for disaster response.	Medium to long term
	Build climate literacy within the CENTCOM command by hiring civilian staff and contractors and working with joint professional military education institutions to expand offerings on this issue.	Short to medium term
	Conduct climate-informed analyses on key installation and logistics routes to understand how climate hazards will influence the local physical and security environment and affect force posture and response.	Medium to long term
Encourage partners and allies to incorporate climate impacts into their force posture and planning in the AOR	Establish bi- and multilateral arenas to share technology to mitigate and adapt to the impact of climate change (e.g., personal cooling equipment, all-weather ISR).	Short to medium term
	Include climate analysis in intelligence exchanges, key leader engagements, and CHOD meetings to improve partner understanding of the real-time effects of climate hazards.	Short to medium term
	Work with DoD to expand and provide DCAT to regional partners to assess installation resilience to climate hazards.	Medium term
	Through civil affairs engagements, identify host country vulnerabilities in responding to climate hazards that could lead to public health crises, acute food and water insecurity, and mass migration. Develop plans to close those gaps.	Short to medium term
	Work with partners to develop defense climate action and implementation plans.	Short term
	Build climate literacy for U.S. military representatives who are responsible for defense relationships abroad, such as security cooperation officials and defense attachés, by expanding training and professional military education.	Short to medium term
	USACE, with U.S. interagency partners, should work with host nations to ensure built and natural infrastructure is climate resilient.	Medium to long term
	Identify with partners potential projects that could be funded under the DORIC program.	Short term
Prepare for an increase in HADR operations, including cooperation with regional partners and affected host nations	Conduct bi- and multilateral regional exercises to improve disaster response and recovery missions, using JTEEP funds when appropriate.	Short term
	Expand the SPP to include National Guard partnerships with CENTCOM countries at high risk for extreme weather events (e.g., Pakistan).	Medium to long term
	Conduct HADR-specific exercises and training with regional partners and U.S. interagency partners (e.g., USAID, FEMA).	Short term
	Engage the Center of Excellence for Disaster Management and Humanitarian Assistance to conduct a climate change assessment on the most likely hazards and locations for future HADR missions and how these findings might influence response force requirements.	Short to medium term
	Update HADR and NEO CONPLANs to include climate projections.	Short term

NOTE: CHOD = Chief of Defense; CONPLAN = contingency plan; DCAT= DoD Climate Assessment Tool; DORIC = Defense Operational Resilience International Cooperation; FEMA = Federal Emergency Management Agency; ISR = intelligence, surveillance, and reconnaissance; JTEEP = Joint Training Exercise and Evaluation Program; NEO = noncombatant evacuation operations; USACE = U.S. Army Corp of Engineers; USAID = U.S. Agency for International Development. Short term is in the next one to three years, medium term is three to seven years, and long term is seven to ten years.

accelerate climate change-related knowledge to and from DoD. In concert with partner nations, ensure overseas infrastructure is adapted and resilient to local conditions.[6]

Among GCCs, CENTCOM has a particularly strong motivation for leveraging climate adaptation to deepen cooperation among partners. Specifically, the 2021 change in the Unified Command Plan that designated Israel as part of the theater is a watershed—yet sensitive—moment in the command's evolution, given long-standing Arab opposition to overt security cooperation with Israel. However, Israel is a global leader in desalination technology, among other climate-adaptation expertise.[7] In the wake of the Unified Command Plan change, Israel, the UAE, and Jordan have expanded collaboration on both renewable energy and desalination to the extent that Jordan swaps renewable energy with Israel to expand its access to water and the UAE provides investment capital.[8]

Taking Cues from Regional Partners

Regional partners are already experiencing, and thus are acutely aware, of the impacts of climate change on their physical environment. Moreover, partners are already responding in a variety of ways to increase their voice and activity related to climate change. CENTCOM partners have become critical participants and hosts of the past two Conference of the Parties of the United Nations Framework Convention on Climate Change, more commonly known as COP. Egypt and the UAE, as the hosts of COP27 and COP28, respectively, fulfill key roles in the negotiations and successes of the summits.[9] During COP27 in 2022, following devastating flooding earlier that year, Pakistan coordinated an effort to establish a Fund for Loss and Damage that drove a conversation between the countries that are most responsible for carbon-based greenhouse gas emissions and the countries that are being affected the most.[10]

In 2023, the UAE seeks to ensure that COP28 focuses on ways to reduce greenhouse gas emissions while the global economy maintains energy security through responsible energy transitions.[11] The focus on renewable and sustainable energy is not new for the UAE and aligns with its *UAE Energy Strategy 2050* (see Box 2.1).[12] The Gulf Cooperation Council countries of the Central Gulf subregion have an opportunity to transition from petroleum-based to more diversified economies that include larger non-energy sectors and renewable energy.[13] All Gulf Cooperation Council members have substantial areas that receive daily, direct, normal irradiation from the sun above 5 kilowatts per hour and areas with wind speeds above 5 meters per second at a height of 50 meters, which economically justify large solar and wind farms, respectively.[14] Saudi Arabia aspires to emerge as the world's leading exporter of hydrogen; in Qatar, land and water management constitute primary focal points in its climate strategy.[15]

Leveraging Innovation Efforts

As some regional countries are already developing their own climate strategies, CENTCOM can capitalize on these initiatives via three potential areas for future cooperation: processes and partnerships, the provision of physical infrastructure and products, and technology transfer. First, CENTCOM can use adaptation efforts as a catalyst to foster regional collaboration among its coalition members, building on the initial efforts of the Abraham Accords. Processes and partnerships encompass lessons and operational efficiencies that DoD and its component services have adopted to reduce energy consumption and emissions and to sequester carbon. Such initiatives as Army Compatible Use Buffer can be adopted by DoD facilities overseas, if feasible, and DoD could share process improvements and efficiencies with suitable partners in CENTCOM to reduce fuel consumption.[16] Next, DoD could find ways to provide partners with certain nonsensitive products, such as surplus fuel-efficient or electric non-tactical and hybrid tactical vehicles. Furthermore, DoD could establish small-scale bio-refineries (once tested and deemed acceptable) in countries where the U.S. military maintains a larger military footprint to source biomass or other components for blended fuel locally or in theater. Finally, CENTCOM can spearhead efforts to facilitate technology transfers and knowledge-sharing via regional forums, where partners with large scientific ecosystems and industrial bases can participate in initiatives to boost supply chain resilience and the adoption of cleaner fuels and technologies.

However, the sensitivity of certain products and technologies could pose barriers to collaboration. Furthermore, many U.S. partners, particularly in the CENTCOM region, might have relationships with global competitors, such as China, or regional adversaries, such as Iran. In that case, the provision of physical products and technology-sharing would require rigorous end-user certifications and assessments to ensure that technology developed by the United States and its allies does not fall into the hands of adversaries.

Box 2.1. United Arab Emirates: Clean Energy Collaboration with the United States

In 2017, the UAE introduced the *UAE Energy Strategy 2050*, the country's first unified approach to addressing energy supply and demand. The strategy aims to increase the share of clean energy in the total energy mix from 25 percent to 50 percent by 2050, reduce the carbon footprint of power generation by 70 percent, and boost individual and corporate consumption efficiency by 40 percent. The targeted energy mix comprises 44 percent clean energy, 38 percent gas, 12 percent clean coal, and 6 percent nuclear power.

The United States and the UAE have been working closely on clean energy initiatives with regular high-level coordination meetings and significant investments. In January 2023, officials announced a $20 billion allocation to fund 15 gigawatts of clean and renewable energy projects in the United States by 2035, which builds on the 2022 UAE-U.S. Partnership for Accelerating Clean Energy that aims to mobilize $100 billion for clean energy globally. The UAE became the first Central Gulf country to commit to net zero emissions by 2050 at COP26 in 2021 and co-launched the Agriculture Innovation Mission for Climate with the United States. The two countries also hosted the first Regional Climate Dialogue in 2021, securing commitments from the Gulf Cooperation Council and other Middle Eastern and North African countries to reduce emissions by 2030. UAE companies Masdar and the Abu Dhabi Investment Authority made significant investments in U.S. renewable energy projects in 2017 and 2019, further solidifying their commitment to a clean energy future.

SOURCES: United Arab Emirates Government, "UAE Energy Strategy 2050," webpage, updated August 10, 2023; Embassy of the United Arab Emirates, Washington, D.C., "A Shared Commitment to Climate Action," webpage, undated-a.

Conclusion

This chapter has explored how CENTCOM can expand on existing OAIs or introduce new OAIs to respond to climate change. We developed these OAIs through three complementary approaches. The first approach was to identify OAIs that could serve as off-ramps from our prior research on causal pathways from climate hazards to conflict. Recognizing that this approach only uncovers OAIs that play a conflict prevention role, we also generated a broader list of OAIs via a structured brainstorm that addressed U.S. and partner force posture and enhanced preparedness for HADR. Finally, we considered OAIs that have an ancillary benefit of deepening the CENTCOM coalition through the sharing of scientific knowledge and technology.

The results of our research suggest that CENTCOM is not the primary U.S. government agency with the mandate and capabilities to break the progression from climate hazards to conflict, but CENTCOM does have an important supporting role to play via military-led OAIs. Specifically, CENTCOM could help interrupt the progression from climate hazards to intrastate conflict by developing partner forces with improved security governance, enhanced disaster response capabilities, and more resilient logistics, among other activities. Since CENTCOM must also adapt to climate stress and support partner adaptations beyond interrupting the climate-conflict nexus, our research also identified a multitude of additional OAIs that CENTCOM could field that would enhance its own and partner preparedness for climate hazards. Finally, our research suggested opportunities for deepening within-CENTCOM cooperation via technology-sharing related to improving fuel efficiency and lowering emissions from military transport; desalination and renewable energy are also key areas for innovation and cooperation.

Endnotes

1 U.S. Central Command, "Operations and Exercises," webpage, undated.

2 Chandler et al., 2023.

3 National Guard Bureau, "National Guard State Partnership Program," map, May 17, 2023. It should be noted that CENTCOM (21 countries) is also a smaller AOR than U.S. Africa Command (AFRICOM) (53 countries), U.S. Indo-Pacific Command (INDOPACOM) (36 countries), U.S. European Command (EUCOM) (51 countries and territories), and U.S. Southern Command (SOUTHCOM) (31 countries).

4 Tom Ellison and Erin Sikorsky, "CCS Releases New Tool: Military Responses to Climate Hazards Tracker," Council on Strategic Risks, June 6, 2023.

5 Júlia Palik, Siri Aas Rustad, Kristian Berg Harpviken, and Fredrik Methi, *Conflict Trends in the Middle East, 1989–2019*, Peace Research Institute Oslo, 2020, p. 9.

6 Office of the Under Secretary of Defense (Acquisition and Sustainment), *Department of Defense Climate Adaptation Plan*, U.S. Department of Defense, September 1, 2021, p. 19.

7 Rowan Jacobsen, "Israel Proves the Desalination Era is Here," *Scientific American*, July 29, 2016.

8 Sue Surkes, "Israel, Jordan, UAE Sign New MOU on Deal to Swap Solar Energy for Desalinated Water," *Times of Israel*, November 8, 2022.

9 Embassy of the United Arab Emirates, Washington, D.C., "COP28," webpage, undated-b.

10 Nina Lakhani, "'We Couldn't Fail Them': How Pakistan's Floods Spurred Fight at COP for Loss and Damage Fund," *The Guardian*, November 20, 2022; Ministry of Foreign Affairs, Government of Pakistan, "Pakistan Welcomes the Historic Decision of COP27 to Establish the Fund for Loss and Damage," press release, November 20, 2022.

11 Embassy of the United Arab Emirates, Washington, D.C., undated-b.

12 Tariq Al Fahaam and Hazem Hussein, "COP28 Will Be the UAE's Most Important Event in 2023: Sheikh Mohammed bin Rashid," Emirates News Agency, November 23, 2022; Chloé Farand, "UAE Plans to Have It Both Ways as COP28 Climate Summit Host," Climate Home News, June 12, 2022.

13 Li-Chen Sim, "Renewable Power Policies in the Arab Gulf States," Middle East Institute, February 8, 2022.

14 Amin Mohseni-Cheraghlou, "Fossil Fuel Subsidies and Renewable Energies in MENA: An Oxymoron?" Middle East Institute, February 23, 2021.

15 Saudi Arabia has adopted five distinct measures to tackle carbon dioxide emissions: (1) reducing, reusing, recycling, and removing carbon dioxide emissions; (2) investing in renewable energy; (3) emerging as the world's top producer and exporter of clean hydrogen; (4) enhancing energy efficiency; and (5) revolutionizing waste management. See Saudi & Middle East Green Initiatives, "SGI Target: Reduce Carbon Emissions by 278 mtpa by 2030," webpage, undated.

Qatar's *National Environment and Climate Change Strategy* prioritizes (1) greenhouse gas emissions and air quality, (2) biodiversity, (3) water, (4) circular economy and waste management, and (5) land use. The country aims to address greenhouse gas emissions and air quality by reducing emissions by 25 percent by 2030, establishing a national air quality monitoring network by 2023, and creating a national air emissions inventory by 2023. Water-related goals include using reverse osmosis and sustainable technologies for more than 55 percent of desalinated water production, reducing groundwater extraction by 60 percent, decreasing per capita household water consumption by 33 percent from 2019 levels, and ensuring 100 percent of recycled water is reused. See Government Communications Office, State of Qatar, "Environment and Sustainability," webpage, undated.

16 The Army has used a system of local partnerships called the Army Compatible Use Buffer to preserve private land near Camp Shelby, Mississippi, and achieve the sequestration of about 120,000 tons of carbon dioxide equivalent annually. See Department of the Army, Office of the Assistant Secretary of the Army for Installations, Energy and Environment, *United States Army Climate Strategy*, February 2022.

CHAPTER 3

MILITARY OPERATIONS AND THEIR COSTS

AS DISCUSSED IN THE previous reports in this series, the connection between climate change and conflict is not necessarily as direct as it is often imagined. Existing empirical studies and our own research suggest that the relationship between climate hazards and large-scale conflict is typically indirect.[1] Specifically, climate hazards interact with other drivers (e.g., economic underdevelopment, poor governance, prior exposure to conflict) to increase the frequency and intensity of armed conflict. Although this body of research should temper defense planners from assuming that resource scarcity will directly translate to conflict, there is evidence that climate exacerbates the conditions that produce conflict and, therefore, can be considered a threat multiplier. There are also reasons to suggest that the CENTCOM AOR could be particularly vulnerable to climate-related conflict because

1. the underlying socioeconomic conditions that climate interacts with to produce conflict are prevalent in the theater
2. the CENTCOM AOR starts from a baseline that already encompasses high levels of extreme heat and water scarcity
3. CENTCOM has already experienced a substantial increase in intrastate conflict, including internationalized civil wars, over a roughly 20-year period beginning in 1996.[2]

Therefore, there might be points in the future when CENTCOM will have to respond to climate-related conflict even if it introduces the off-ramps addressed in Chapter 2.

To assist CENTCOM planners as they consider how best to meet future requirements in a climate-affected security environment, we discuss interrelated topics concerning the use of force. We begin by analyzing historic patterns and indicators of the use of military force by the United States. We couple this analysis with ROM cost estimates for four operation types (counterterrorism, stabilization, HADR, and NEOs) because defense planning is obliged to match requirements to the availability of forces and resources.

Past Patterns of U.S. Military Interventions

The potential for climate hazards to correspond with increases in armed conflict in the CENTCOM AOR is, from a policymaker perspective, only half the story. Understanding the historical patterns of how and why the United States engages militarily can guide planners as they prepare for future contingencies. RAND researchers have conducted a series of projects that analyze historic trends in U.S. military interventions and have identified factors that influence the likelihood and size of these interventions.[3] This research has resulted in several reports and the production of an original dataset of global U.S. military interventions, the RAND U.S. Military Intervention Dataset (RUMID), covering 1898 through 2017, which also includes qualitative case studies of several U.S. interventions in this period. The threshold for inclusion in the dataset was "any deployment of U.S. ground troops to another sovereign country that included at least 100 'person years'" (i.e., could be 100 troops deployed for one year or a larger number deployed for a shorter amount of time) and larger threshold sizes for naval and air interventions.[4]

RAND researchers identified four types of U.S. military intervention, including ground, air, and naval intervention, that were most common in this period: interventions into armed conflicts, deterrence missions, stability operations, and HADR.[5] The data reveal that, globally, deterrence operations were the dominant intervention type during the Cold War. However, since 1990, humanitarian interventions have become the most numerous type, exceeding combat interventions and stability operations combined.[6] Figure 3.1 illustrates the change over time in the balance of interventions.

RUMID undercounts the true number of HADR operations because of the force size thresholds it uses to determine what qualifies as an intervention.[7]

To understand how frequent this operation type occurs in the CENTCOM AOR, Figure 3.2 presents the geographic clustering of HADR operations that the U.S. military has conducted in the region from 1970 to mid-2023. The numbers that appear next to the country name and the size of the corresponding bubble indicate the frequency of HADR operations in the affected country. A single country within the 21-country AOR—Pakistan—was the site of more than one-third of total HADR operations (12 of 32) during that half century. Pakistan is also projected to receive even more-extreme weather events in the future under the climate projections documented in the first report in this series.[8]

Figure 3.1. U.S. Global Interventions by Activity Type, 1898 to 2017

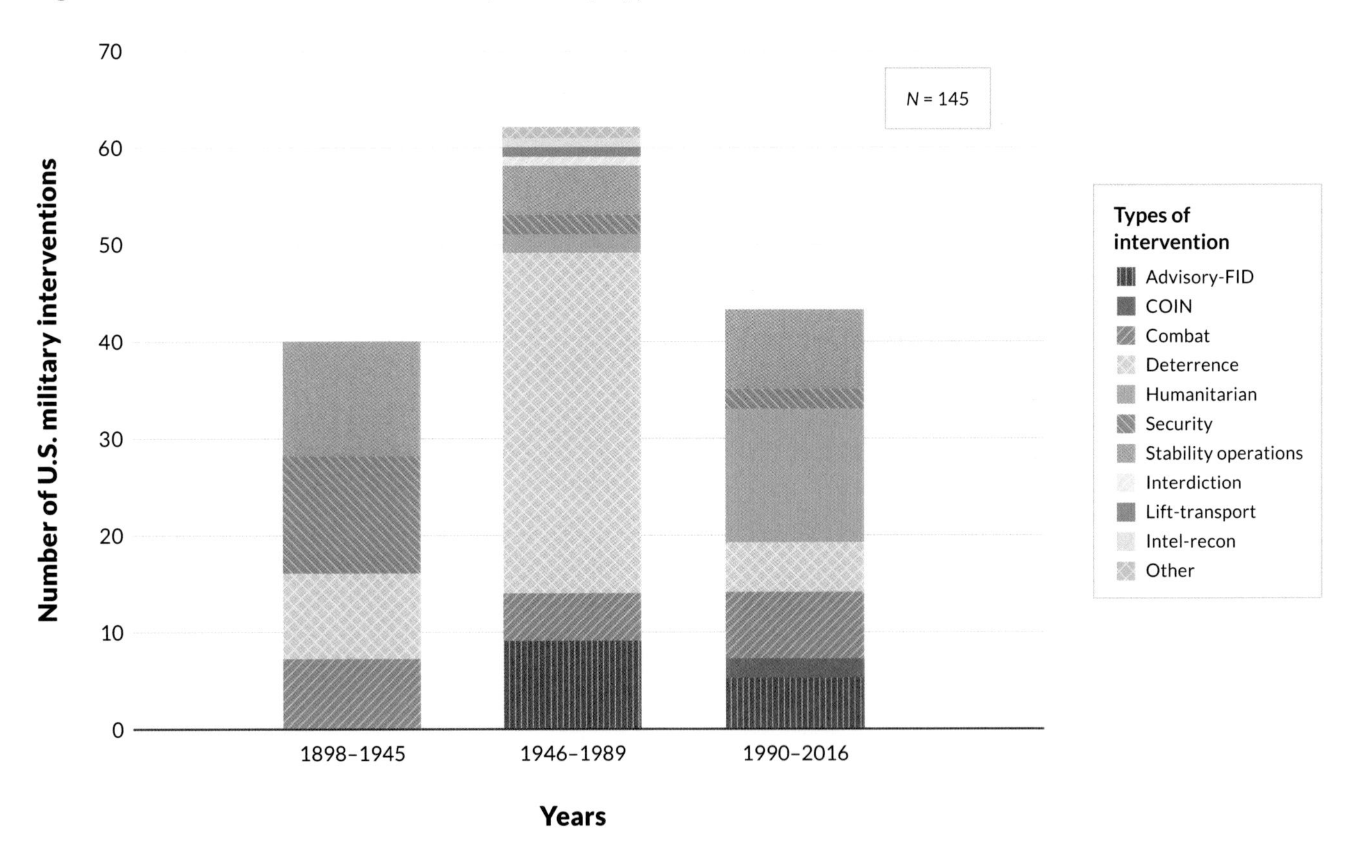

SOURCE: Adapted from Jennifer Kavanagh, Bryan Frederick, Alexandra Stark, Nathan Chandler, Meagan L. Smith, Matthew Povlock, Lynn E. Davis, and Edward Geist, *Characterizations of Successful U.S. Military Interventions*, RAND Corporation, RR-3062-A, 2019.

NOTE: COIN = counterinsurgency; FID = foreign internal defense; intel-recon = intelligence and reconnaissance.

Figure 3.2. Geographic Breakdown of Humanitarian Assistance and Disaster Response in Central Command

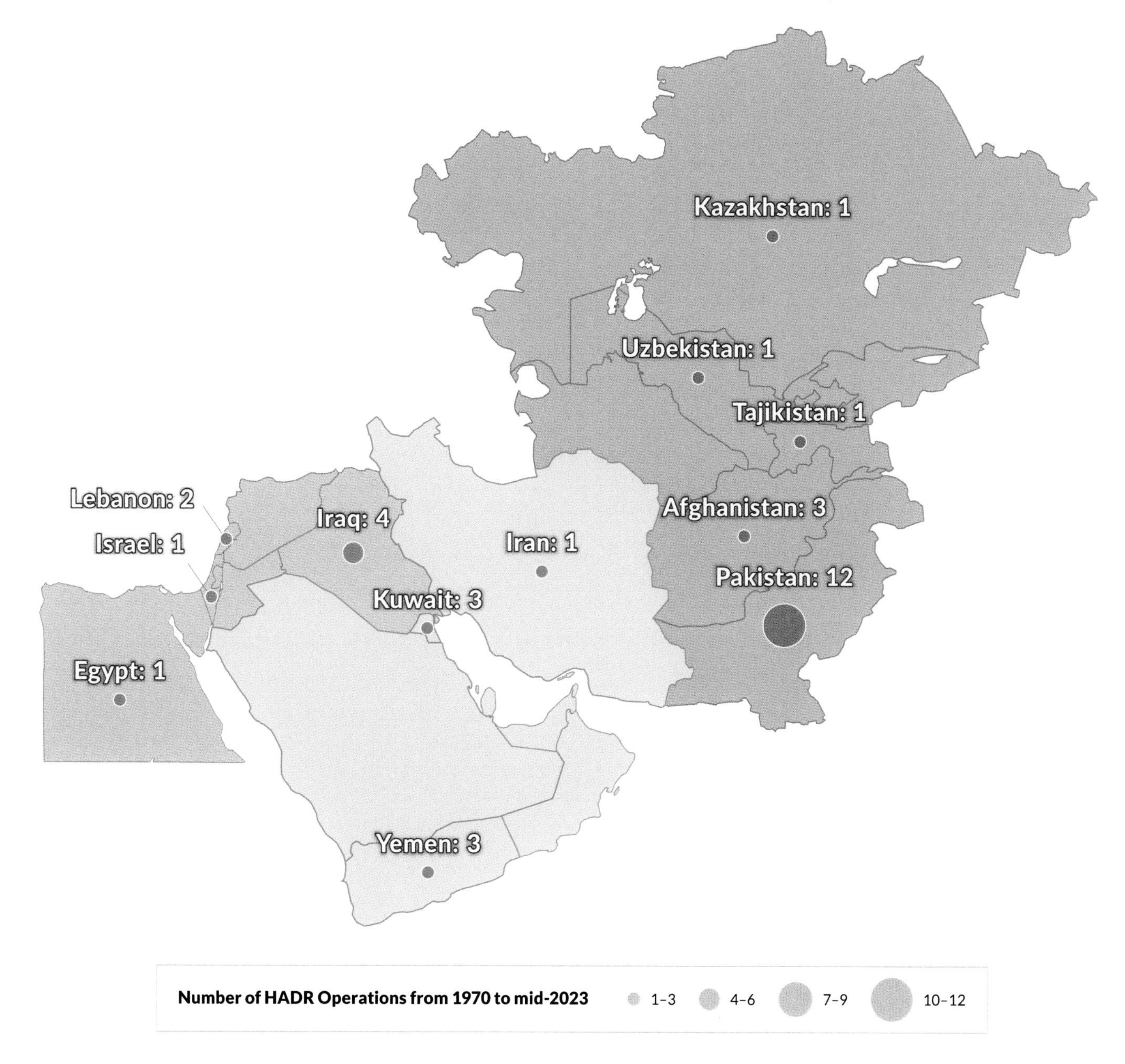

SOURCES: The 32 HADR operations depicted in the figure are documented in W. Eugene Cobble, H. H. Ganey, and Dmitry Gorenburg, *For the Record: All U.S. Forces' Responses to Situations, 1970–2000 (with Additions Covering 2000–2003)*, CNA Center for Strategic Studies, May 2005. We augmented these data by searching USAID fact sheets dating back to 2000 that included any mention of DoD contributions.

NOTE: Because one HADR mission spanned both Pakistan and Afghanistan, the numbers in the figure sum to 33 rather than 32 total operations. The numbers also exclude HADR operations conducted within prior CENTCOM borders, such as when the command encompassed Kenya and Somalia.

Signposts and Costs of Interventions by Type

To support military planners, we complement our analysis of the historic frequency of U.S. military interventions with a review of the conditions that make the intervention type more likely, which we refer to as *signposts*, and ROM costs for each intervention type. We focus on four types of DoD interventions and activities: counterterrorism operations, stability operations, HADR, and NEOs. We provide ROM costing analysis on security cooperation activities in the appendix. We treat these intervention types as distinct but acknowledge that there could be overlap across these categories (e.g., a stability operation could include a HADR element). Our analysis of signposts is derived from prior RAND studies on military interventions referenced previously in this chapter. Our costing analysis is derived from original research and does not seek to provide point estimates of the cost of these interventions, given that they encompass large variations in costs depending on the size and length of the intervention. Instead, we present the range of historic costs as ROM estimates only. For the costing component, we focus on DoD activities that were carried out in the past 20–25 years under the assumption that more recent cases are more applicable to the 2035–2070 horizon that is guiding this study.

The type of costing information available varies significantly across the intervention types. In some cases, we were able to find historic cost data from official government sources. In other cases, the costs were not readily available. When information on costs was not readily available, we either modeled the costs from open-source information or described the forces deployed to provide a sense of the magnitude of the intervention. In short, the cases and associated costs outlined in the following paragraphs are intended to be illustrative rather than fully representative and generalizable. Our analysis indicated that there is significant variation in costs between and within each intervention type. In general, the highest cost interventions and activities tended to be those that are associated with stability operations. In contrast, HADR, counterterrorism, and NEOs were less costly in terms of the resources and forces deployed.

Armed Conflict Summary

As mentioned earlier, Kavanagh et al. (2017, p. 195) defines *armed conflict* as "[i]nterventions involving traditional military operations and fighting, characterized by large formations of organized military forces on both sides." Although armed conflicts fitting this traditional definition do not occur often, the Middle East in particular has been the scene of many such conflicts in the second half of the 20th century. Hence, understanding how and when the United States might militarily engage in armed conflicts, which can also be referred to as major combat operations (MCOs), is important for CENTCOM planners. We additionally did not cost MCOs because of their scope.

According to RUMID, the United States intervened in about 20 percent of all armed conflicts from 1898 through 2015. The percentage of intervention in ongoing conflicts in a given year has evolved over time, however, reaching a peak of 50 percent of all armed conflicts in 2006 before declining.[9] Compared with the other types of intervention considered here, combat operations tend to be larger than other types of intervention but also shorter in duration.[10]

The factors that affected the likelihood and size of interventions into armed conflict included the following:

- destructiveness of the war
- previous intervention
- war weariness
- relationship with the United States
- U.S. capabilities
- elite and public opinion
- attack on U.S. soil.

The existence of a previous intervention in the same country makes a new combat intervention more likely. In the third report in our series, we ran models to project the incidence of conflict in the CENTCOM AOR through 2070 under different climate and socioeconomic scenarios. These models suggest that we could see an increased risk of conflict hot spots, including in countries where the United States has intervened in recent decades, such as Afghanistan, Iraq, Pakistan (in a counterterrorism operation), and Syria.[11] If climate hazards increase the risk of conflict in these hot spots, we could also see an increased likelihood of recurring interventions in conflicts that take place in those locations.

The risk of intervention increases if armed conflict takes place within the territory of a U.S. ally. However, these types of interventions tend to be smaller in size, possibly because U.S. allies could be more capable and therefore less in need of external support. Additionally, U.S. forces tend to get involved at the earlier stages of a crisis in ally countries compared with conflicts in other countries.[12] The closeness of the United States' relationship with allies and partners in the CENTCOM

AOR is long-standing and not directly tied to climate change, but these relationships nevertheless evolve over time.

Counterterrorism Operations

We have included counterterrorism operations—and not MCOs, such as the 1990–1991 Gulf War—in our costing analysis because the CENTCOM AOR remains a key source of terrorist threats; CENTCOM defines this mission as one of three core lines of effort, with deterring Iran and strategic competition consisting of the other two lines of effort. Moreover, we found more evidence in our research for climate stress contributing to conflict below the threshold of war and intrastate war than we did for climate stress contributing to interstate war.[13]

There are various vulnerabilities related to climate change in the AOR that terrorist groups can exploit. As extreme heat and decreasing water supplies threaten economic livelihoods and food security, terrorist organizations can leverage these conditions to recruit new members, especially among internal migrants who are displaced to urban centers by climate change.[14] State actors could also be weakened by climate change (i.e., stretched thin by disaster relief, the loss of revenue from declining agricultural sectors) and less able to contain terrorist groups.[15] Moreover, some terrorist groups—traditionally those in Latin America—have environmentally focused agendas around combatting land appropriation that might be further inflamed by climate change.[16] Finally, terrorist groups could try to manipulate the physical environment stressed by climate to advance their aims. As a CENTCOM-specific example, the Islamic State of Iraq and Syria used its control of dams as a coercive tool to restrict water access to downstream communities and to provide more water for agrarian communities under its control.[17]

Counterterrorism operations can involve any number of instruments, varying from security cooperation, to air strikes, to ground operations. For two reasons, we focus on estimating the costs associated with one of these instruments: targeted drone campaigns. First, recent trends in national security strategy have relegated counterterrorism as a mission priority, generating more attraction to drone strikes as a less costly alternative to large-scale ground operations or prolonged air campaigns using manned platforms.[18] Second, for this costing exercise, we have endeavored to select conservative samples of OAIs to cost, which is in line with our central finding across the reports in this series that climate is a threat multiplier rather than the central driver of current and future conflict.

Because of the challenges calculating the actual operational costs of drone campaigns, we employ a cost model derived from the largest direct cost variables that can be identified from open-source information. Within this cost model, we focus specifically on the operational costs of drone campaigns, such as operations and maintenance (O&M) flight costs and munitions costs, as opposed to personnel or fixed capital costs.[19] Table 3.1 only includes the total operational costs for our four cases (Somalia, Libya, Yemen, and Pakistan); a more detailed analysis of the costs that make up these totals can be found in the appendix.

Stabilization Operations

With the withdrawal or reduction of U.S. military forces in Afghanistan and Iraq, respectively, CENTCOM has lower involvement in stabilization operations than during the height of the Global War on Terror. However, despite this troop reduction and a decreased interest in stabilization operations among U.S. policymakers, CENTCOM does remain engaged in smaller-scale stability operations in the AOR and will likely remain so for the foreseeable future. Stability operations in RUMID include nation-building, peacekeeping, and other law and order–focused tasks. Most stability operations in the dataset occur in conflict or post-conflict settings within ten years of an armed conflict's end.

The factors that affect the likelihood and size of stability operations include the following:

- number of refugees
- involvement in the combat phase of an armed conflict
- location (region) of the target
- wealth of the target nation
- military assistance
- presence of a multilateral coalition.

RAND researchers have found that a large number of refugees (and, by extension, the presence of a humanitarian crisis) can be a driver for U.S. involvement in stability operations. Migration can result from any number of precipitating factors, however, and climate-related events are one contributing factor to temporary and permanent human migration patterns. A future in which there are a larger number of refugees because of climate change could also feature an increase in the number of stability operations. One important caveat is that a substantial number of refugees is required before this factor on its own makes intervention significantly more likely, and a large number of refugees corresponds with a relatively small increase in the probability of intervention in RAND's intervention

Table 3.1. Modeled Historic Costs of Selected Targeted Drone Campaigns, 2004 to 2019

	Somalia 2011–2018	Libya 2012–2019	Yemen 2009–2013	Pakistan 2004–2018
Total operational costs (annual)	$9.7–21.9 million	$14.7–33 million	$10.1–22.3 million	$14.9–33.2 million
Total operational costs (all years)	$77.7–175 million	$117.4–263.9 million	$50.5–111.5 million	$223–497.6 million

SOURCE: Peter Bergen, David Sterman, and Melissa Salyk-Virk, "America's Counterterrorism Wars: Tracking the United States's Drone Strikes and Other Operations in Pakistan, Yemen, Somalia, and Libya," New America, June 17, 2021; Jack Serle and Jessica Purkiss, "Drone Wars: The Full Data," database, Bureau of Investigative Journalism, January 1, 2017; Nick Turse, Henrik Moltke, and Alice Speri, "Secret War," The Intercept, June 20, 2018; Anne J. McAndrew, "Fiscal Year (FY) 2019 Department of Defense (DoD) Fixed Wing and Helicopter Reimbursement Rates," memorandum to assistant secretaries, deputy chiefs, directors of the military departments, et al., Office of the Under Secretary of Defense (Comptroller), October 12, 2018; Cory T. Anderson, Dave Blair, Mike Byrnes, Joe Chapa, Amanda Collazzo, Scott Cuomo, Olivia Garard, Ariel M. Schuetz, and Scott Vanoort, "Trust, Troops, and Reapers: Getting 'Drone' Research Right," War on the Rocks, April 3, 2018.

NOTE: The time frame of each campaign is based primarily on the years for which data on air strikes are available. For Yemen, we focused on the period between 2009 and 2013, after which the United States' involvement shifted from a counterterrorism campaign to an intervention in a civil war.

models.[20] This suggests that even if climate-related events result in an increased number of humanitarian crises and refugees, the number of U.S. stability operations in response might not increase in a linear way.

RAND researchers have also found that stability operations are heavily shaped by the region where they take place. The United States is about twice as likely to conduct stability operations in the Middle East than in other regions (it is also more likely to conduct stability operations in Europe).[21] Stability operations are also more likely when the United States is involved in the combat phase of an intervention.[22] If climate-related factors increase the incidence of interventions in armed combat, we might also expect to see increased U.S. involvement in stability operations.

Stabilization Costing

To identify the historical costs of stabilization, we split stabilization cases into two categories: one for North Atlantic Treaty Organization (NATO) operations, for which the DoD contribution can be directly estimated, and the (more common) shared United Nations (UN) stabilization missions, in which the United States has historically played a key role. Figure 3.3 displays the overall costs associated with the NATO and UN stabilization missions included in our sample and the estimated U.S. or actual DoD contributions to those missions. For the UN cases, we chose to include all cases listed on the UN Peacekeeping website that ended after 2000; however, because of the lack of cost data, we excluded operations that ended on or after 2017. We specifically did not address operations in Afghanistan and Iraq during the first two decades of the 21st century; because of their size and scope, those operations would be anomalies for ROM costing estimates.

Previous RAND research found that U.S. contributions to UN peacekeeping missions account for up to 28 percent of total UN mission costs.[23] We used this average to estimate U.S. contributions to the UN missions. We acknowledge that the U.S. funds for UN peacekeeping are almost entirely from departments other than DoD, but these cases can be helpful bellwethers of potential DoD costs. Specifically, the U.S. Government Accountability Office has put out several reports that estimate that a unilateral U.S. state-building effort would greatly exceed the cost of a comparable UN operation.[24] The appendix includes additional information on the NATO and UN missions included in our sample, including a brief description of each mission. As with the other operation types, DoD and U.S. costs associated with stability operations ranged significantly from about $5.9 million (UN Supervision Mission in Syria, 2012) to $18.9 billion (Bosnia, 1995).

Figure 3.3. U.S. Estimated Contributions to UN and NATO Peacekeeping and Stabilization Missions, 1999–2012

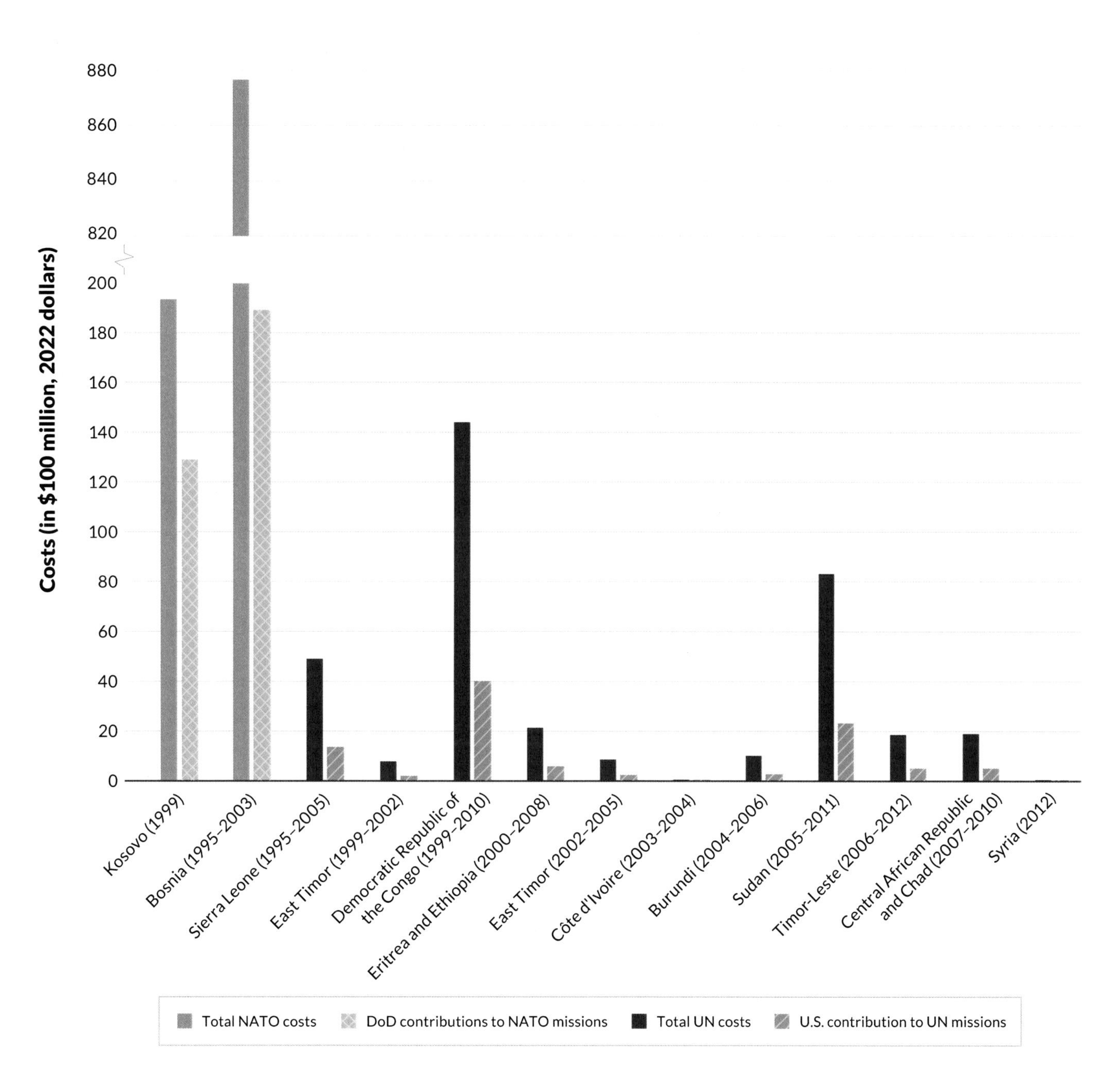

SOURCES: UN Integrated Mission in Timor-Leste, "UNMIT Facts and Figures," fact sheet, undated; UN Mission in Côte d'Ivoire, "Côte d'Ivoire—MINUCI—Facts and Figures," fact sheet, United Nations Peacekeeping, 2004; UN Mission in Ethiopia and Eritrea, "Ethiopia and Eritrea—UNMEE—Facts and Figures," fact sheet, United Nations Peacekeeping, 2009; UN Mission in Sierra Leone, "Sierra Leone—UNAMSIL—Facts and Figures," fact sheet, 2005; UN Mission in the Central African Republic and Chad, "MINURCAT Facts and Figures," fact sheet, United Nations Peacekeeping, undated; UN Mission in the Sudan, "UNMIS Facts and Figures," fact sheet, undated; UN Mission of Support in East Timor, "East Timor—UNMISET—Facts and Figures," fact sheet, United Nations Peacekeeping, 2005; UN Operation in Burundi, "Burundi—ONUB—Facts and Figures," fact sheet, United Nations Peacekeeping, 2007; UN Organization Mission in the Democratic Republic of the Congo, "MONUC Facts and Figures," fact sheet, United Nations Peacekeeping, undated; UN Transitional Administration in East Timor, "East Timor—UNTAET: Facts and Figures," fact sheet, United Nations Peacekeeping, 2002; UN Supervision Mission in Syria, "UNSMIS Facts and Figures," fact sheet, undated; Martin Sletzinger, "Iraq Through the Lens of Bosnia and Kosovo," Wilson Center, March 17, 2003; Benjamin Zyla, *Sharing the Burden? NATO and Its Second-Tier Powers*, University of Toronto Press, 2015.

NOTE: U.S. contributions were estimated by taking 28 percent of total UN peacekeeping costs. Although virtually none of the funding came from DoD, these numbers reflect the lower bound costs that DoD might incur for such operations if it were to conduct stability operations unilaterally or in coordination with NATO.

Humanitarian Assistance and Disaster Response

Previous analyses from RAND did not explicitly assess signposts for HADR and NEOs. As a result, we do not have a similar set of prior models that identify the factors that make U.S. HADR interventions or NEOs more likely. (A more detailed discussion of NEOs follows in the section "Noncombatant Evacuation Operations.") To address the lack of HADR signposts in RUMID, we collected a sample of HADR interventions in the CENTCOM AOR from 1970 through 2023 that can provide some preliminary insights into the historical patterns of U.S. HADR interventions in the region.[25] This type of intervention could become more important, if not more prevalent, than it has been historically as the risk of climate-related humanitarian crises and natural disasters increases.

Of the 26 HADR interventions we documented in this sample, most were small compared with other types of interventions in terms of size and duration. Fifteen of these HADR interventions lasted for one month or less (and some for only one day).[26] The short HADR interventions typically involved a small number of U.S. Air Force assets.[27] This suggests that if future U.S. HADR interventions increase in frequency but are of similar intensity, military planners will not need to plan for the involvement of a significantly larger number of forces or assets. However, with a potential increase in frequency, military planners might consider investing in HADR-specific capabilities, including those of partners and allies, to carry out such relatively small HADR interventions more frequently.

We also documented four HADR interventions that lasted for more than a year and were large in scale. Among these interventions, three—Afghanistan from 2001 through 2021 and Iraq from 2003 to 2011 and from 2014 to the present—accompanied large U.S. combat interventions; a fourth intervention—relief to Afghanistan from 1986 to 1993—took place while the United States was providing support for proxy forces on the ground. In addition to these interventions in Afghanistan and Iraq, the United States from 2003 to 2023 was engaged in HADR interventions in Iran (earthquake relief in 2004), Pakistan (disaster relief in 2005–2006, 2009, 2010, and 2022), and Lebanon (sponsorship of unexploded ordnance clearing and demining and the evacuation of 15,000 U.S. citizens). Overall, this pattern suggests that the United States has engaged in HADR interventions alongside large-scale combat interventions and in strategically important countries. Moreover, as local emergency response capabilities are stretched by the frequency and intensity of climate disasters, military forces are frequently called on to support response efforts. From mid-2022 to mid-2023, military forces in dozens of countries responded to almost 100 climate disasters.[28] Planners could therefore focus at least in part on developing HADR capabilities in strategically important areas to decrease response and deployment time frames. Box 3.1 elaborates on how DoD becomes engaged in HADR interventions.

Humanitarian Assistance and Disaster Response Costing

As previously described, U.S. HADR interventions could become more important as a result of climate change, including for CENTCOM. Figure 3.4 presents the costs (both DoD-only costs and total U.S. government costs) that are associated with a sample of U.S. HADR interventions from fiscal year (FY) 1998 to FY 2022. These cases were identified through a review of situation reports on humanitarian crises published by USAID and of prior RAND research. Our initial sample consisted of about 60 U.S. HADR interventions. We restricted this sample to CENTCOM cases, excluding cases involving earthquakes, which are not correlated with climate change; complex humanitarian emergencies stemming from conflict regardless of the combatant command in which they occurred; and cases of disease outbreak regardless of the combatant command in which they occurred. We selected these cases because of our specific focus on the CENTCOM AOR and the intersection between climate change and conflict and because of the underappreciated risk climate change poses to public health. Incorporating complex emergencies stemming from conflict and disease response required us to go outside CENTCOM for sufficient examples.

As shown in Figure 3.5, DoD contributions varied significantly among the HADR operations in our sample, from $296,000 for DoD's response to a 2009 influenza outbreak in Central America to $766 million for its response to the 2014 Ebola outbreak in West Africa. In part, this has to do with the support DoD was requested to provide, which varied significantly across interventions. DoD's response to the 2014 Ebola outbreak, for example, involved a massive effort to design and build Ebola treatment units, medical laboratories, and other medical facilities. For other HADR interventions, DoD assistance was much narrower in scope and included air-dropping daily rations and transporting other relief supplies, such as personal protection equipment or winter clothing. The appendix presents additional information on the cases that were included in our cost analysis and the types of assistance DoD provided for each, and Box 3.2 expands on the discussion of these types of assistance.

Noncombatant Evacuation Operations

Although not specifically discussed as an intervention in RUMID, NEOs—whereby DoS, sometimes in coordination with DoD, organizes the protection and evacuation of embassy staff, dependents, and/or private U.S. citizens—could be required at a higher frequency in the future because of climate-related instability. These evacuations can occur in response to a wide variety of threats, many with potential links to climate change, such as intrastate conflict, terrorist attacks, natural disasters, and disease outbreaks.[29] NEOs can vary from straightforward departures on regularly scheduled commercial flights coordinated by DoS to large-scale sea and air lift operations that require U.S. military ships and planes.[30] A 2007 report from the U.S. Government Accountability Office found that, of the 89 evacuations ordered between 2002 and 2007, almost half were concentrated in the Middle East, Turkey, and Pakistan.[31]

Box 3.1. How Does the Department of Defense Become Involved in Humanitarian Assistance and Disaster Response Missions?

As with other types of interventions, the decision for the United States to become involved in a HADR mission is a policy decision. Unlike some interventions, however, the United States becomes involved in HADR activities when the host country specifically requests such assistance. Once it is determined that the crisis exceeds the host country's ability to respond and the United States has an interest in assisting, the Department of State (DoS) or USAID takes the lead in meeting the humanitarian need. If DoS or USAID require additional USG support, DoD has unique capabilities to contribute. If DoD's support is feasible and appropriate, DoD will respond once it has been approved to do so by the Secretary of Defense. Strategic-level interagency coordination is led by USAID via the director of the Office of U.S. Foreign Disaster Assistance or, for certain large-scale disasters, by the USAID Administrator and the National Security Council, e.g., through an interagency policy committee.

The Office of U.S. Foreign Disaster Assistance may deploy a disaster assistance response team (DART) to assist with coordinating the effort on the ground. DARTs include specialists with relevant disaster relief skills and expertise. In addition to coordinating closely with the U.S. military during foreign disaster relief operations, DARTs assist the U.S. embassy and USAID mission, report on the disaster situation and recommend USG response, and coordinate with nongovernmental and international organizations.

HADR activities conducted by U.S. armed forces, or what DoD calls *foreign humanitarian assistance*, include foreign disaster relief missions (providing basic services and commodities to alleviate suffering, such as supporting necessary logistics and critical infrastructure), dislocated civilian support missions (providing humanitarian support to evacuees, migrants, refugees, internally displaced persons, and stateless persons), security missions (establishing and maintaining the conditions for the delivery of humanitarian relief supplies), and technical assistance and support functions (supporting such tasks as port operations, relief supply distribution and management, and communications restoration).

SOURCE: Joint Publication 3-29, *Foreign Humanitarian Assistance*, Joint Chiefs of Staff, May 14, 2019, pp. I-7, I-8, II-16, II-19, II-20.

Figure 3.4. Department of Defense and Total U.S. Government Costs for Selected Humanitarian Assistance and Disaster Response Operations, Fiscal Year 1998 to Fiscal Year 2022

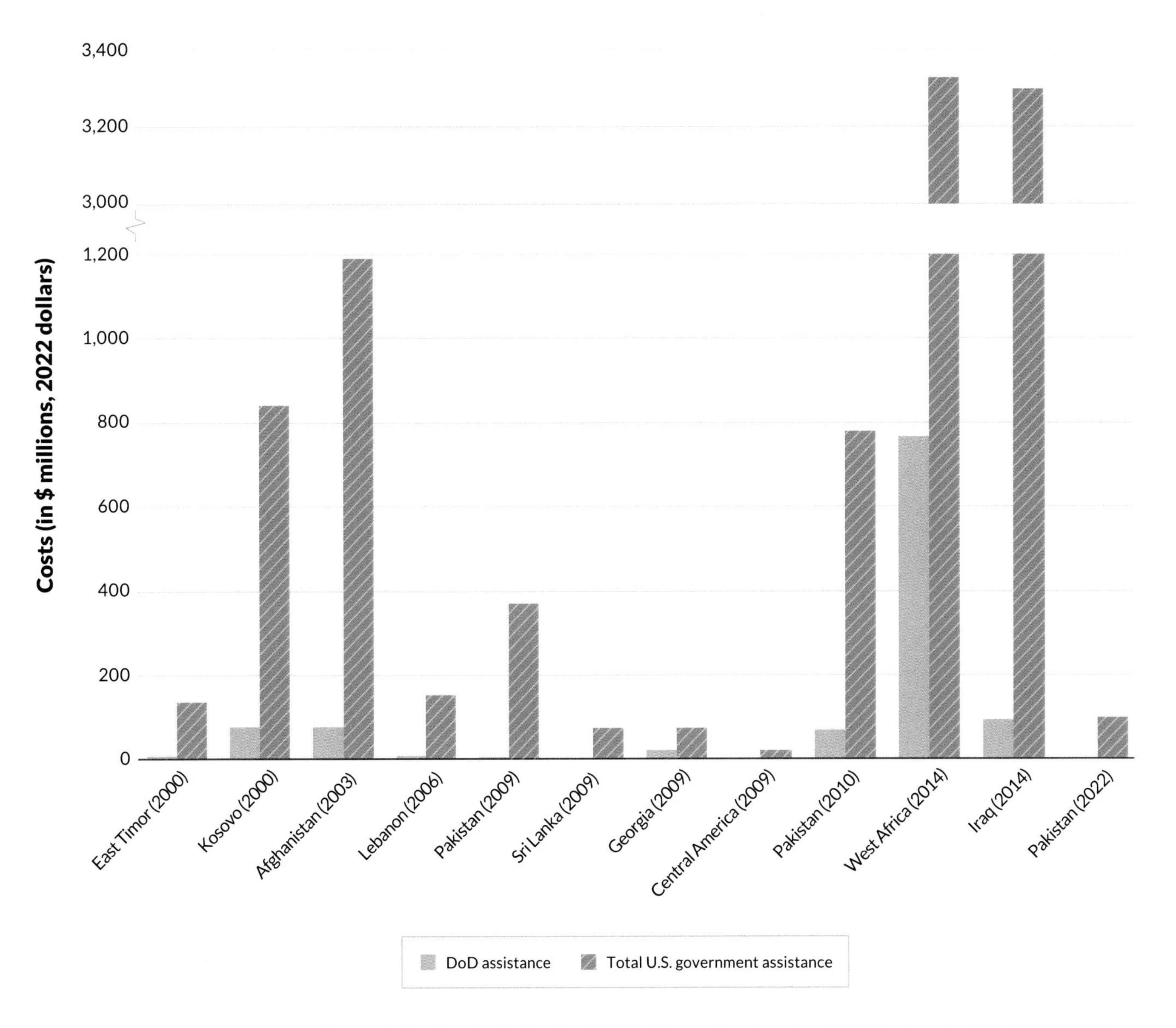

SOURCES: USAID, "Kosovo Crisis Fact Sheet #22," fact sheet, April 26, 1999a; USAID, "Kosovo Crisis Fact Sheet #37," fact sheet, April 11, 1999b; USAID, "Kosovo Crisis Fact Sheet #144," fact sheet, April 7, 2000a; USAID, "East Timor—Crisis Fact Sheet Summary #1, Fiscal Year (FY) 2000," fact sheet, October 3, 2000b; USAID, "Afghanistan Complex Emergency Situation Report #04 (FY 2003)," March 3, 2003; USAID, "Lebanon: Complex Emergency Information Bulletin #1 (FY 2006)," July 21, 2006; USAID, "Sri Lanka—Complex Emergency Fact Sheet #1, Fiscal Year (FY) 2007," fact sheet, February 2, 2007a; USAID, "Lebanon Humanitarian Emergency: USG Humanitarian Situation Report #11 (FY) 2007," June 20, 2007b; USAID, "Global—Influenza A/H1N1 Fact Sheet #1, Fiscal Year (FY) 2009," fact sheet, May 5, 2009a; USAID, "Global—Influenza A/H1N1 Fact Sheet #3, Fiscal Year (FY) 2009," fact sheet, May 18, 2009b; USAID, "Georgia: Complex Emergency Fact Sheet #2, Fiscal Year (FY) 2009," fact sheet, June 18, 2009c; USAID, "Pakistan—Complex Emergency Fact Sheet #31, Fiscal Year (FY) 2009," fact sheet, September 22, 2009d; USAID, "Sri Lanka—Complex Emergency Fact Sheet #18, Fiscal Year (FY) 2009," fact sheet, September 30, 2009e; USAID, "Iraq—Complex Emergency Fact Sheet #1, Fiscal Year (FY) 2012," fact sheet, October 21, 2011; USAID, "Iraq—Complex Emergency Fact Sheet #2, Fiscal Year (FY) 2020," fact sheet, May 8, 2020; USAID, "Central Asia Region Complex Emergency Situation Report #41 (FY 2002)," August 16, 2022; USAID, "Pakistan—Floods Factsheet #2, Fiscal Year (FY) 2023," fact sheet, January 12, 2023; Jennifer D. P. Moroney, Stephanie Pezard, Laurel E. Miller, Jeffrey Engstrom, and Abby Doll, *Lessons from Department of Defense Disaster Relief Efforts in the Asia-Pacific Region*, RAND Corporation, RR-146-OSD, 2013.

Figure 3.5. Distribution of Department of Defense and Total U.S. Government Costs for Selected Humanitarian Assistance and Disaster Response Operations, Fiscal Year 1998 to Fiscal Year 2022 (in 2022 U.S. dollars)

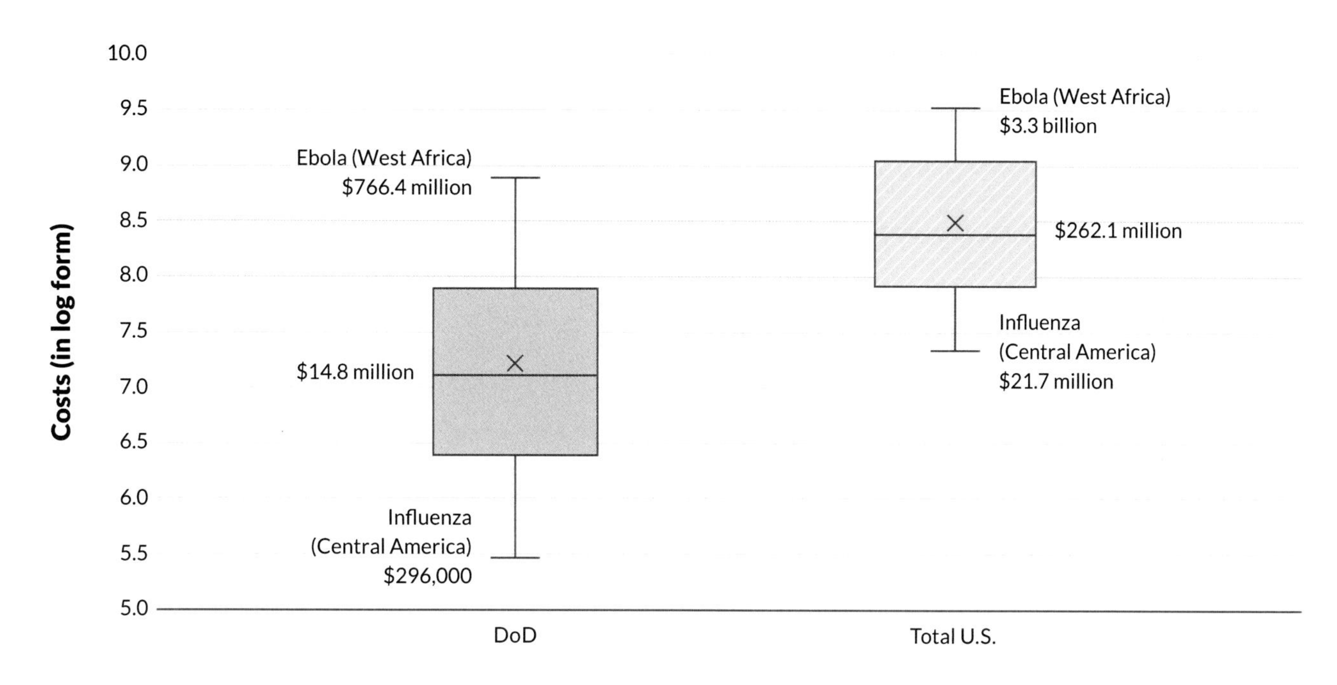

SOURCES: USAID, 1999a; USAID, 1999b; USAID, 2000a; USAID, 2000b; USAID, 2003; USAID, 2006; USAID, 2007a; USAID, 2007b; USAID, 2009a; USAID, 2009b; USAID, 2009c; USAID, 2009d; USAID, 2009e; USAID, 2011; USAID, 2020; USAID, 2022; USAID, 2023; Moroney et al., 2013.

MILITARY OPERATIONS AND THEIR COSTS

We reviewed DoD sources, prior RAND studies, and news articles to develop a list of NEOs conducted by DoD at the request of DoS since 2000. These operations are presented in Table 3.2. We were unable to identify the cost of these operations from open-source materials but include a narrative description of the forces employed to provide a sense of the magnitude of these evacuations. In our sample of cases, the forces employed varied significantly, from the deployment of multiple civilian and military ships to evacuate thousands of U.S. citizens (Lebanon, 2006) to much smaller evacuation operations involving the transport of several hundred people.

To summarize, we examined how and when the United States engaged militarily in armed conflicts and stabilization operations. To provide military planners with an idea of the variety of resources required, we conducted a ROM costing analysis of specific interventions and activities: counterterrorism operations, stabilization operations, HADR, and NEOs. This analysis should assist CENTCOM in understanding how the command could be called on to respond to climate-related conflict in the coming decades.

Table 3.2. Examples of Noncombatant Evacuation Operation Requirements, 2003 to 2023

Case	Short Description	Forces and Systems Employed
Operation Shining Express (Liberia, 2003)	Thirty-five combat-equipped U.S. military personnel were deployed to Monrovia, Liberia, to augment U.S. embassy security forces and aid in the possible evacuation of U.S. citizens.	• USS *Kearsarge* • Three HH-60 Pave Hawk helicopters and their crews • One MC-130P Combat Shadow plane
Lebanon (2006)	DoD evacuated close to 15,000 U.S. citizens from a war zone in less than a month. At least 2,200 marines and sailors in the 24th Marine Expeditionary Unit participated in the evacuation effort.	• Four amphibious ships • Landing Craft Utility boats • Marine Corps CH-53 Super Stallion helicopters
South Sudan (2013)	Forty-five combat-equipped U.S. military personnel evacuated 120 personnel by air from the U.S. embassy in Juba, South Sudan, to Nairobi, Kenya. An additional 46 U.S. military personnel were deployed to Bor, South Sudan, to evacuate U.S. citizens and personnel residing there.	• Two U.S. C-130 aircraft • Three CV-22 Ospreys
Libya (2014)	About eight diplomats and 200 security personnel were evacuated from Tripoli, Libya, to Tunisia by ground convoy. The U.S. Air Force and Marine Corps provided air support for the convoy as a precaution.	• Ground convoy • Three F-17 aircraft • One CV-22 Osprey
Hurricane Irma (St. Maarten, 2017)	Three National Guard units from New York, Puerto Rico, and Kentucky evacuated 1,200 Americans from St. Maarten, flying them to safety in Puerto Rico over the course of two days.	• Six C-130 aircraft • Two HH-60 helicopters
Sudan (2023)	Approximately 1,400 Americans were evacuated over the course of a few weeks by either U.S. forces or U.S. allies and partners.	• Three Chinook Helicopters • USNS *Brunswick* • USS *Truxtin* • USS *Puller*

SOURCES: Scott Schonauer, "Troops Boost Security at African Hot Spots," *Stars and Stripes*, June 14, 2003; Marni McEntee, "USAFE Sends Airmen to Aid Liberia Mission," *Stars and Stripes*, July 15, 2003; Barbara Salazar Torreon and Sofia Plagakis, *Instances of Use of United States Armed Forces Abroad, 1798–2023*, Congressional Research Service, R42738, version 39, March 8, 2022, updated June 7, 2023; Jeff Schogol, "About 1,200 Americans Evacuated from Lebanon," *Stars and Stripes*, July 20, 2006; "11,700 Americans Evacuated from Lebanon," NBC News, July 24, 2006; Jess T. Ford, *State Department: The July 2006 Evacuation of American Citizens from Lebanon*, Government Accountability Office, GAO-07-893R, June 7, 2007a; Patrick Markey, "U.S. Evacuates Libya Embassy After 'Free-Wheeling Militia Violence,'" Reuters, July 27, 2014; Luis Martinez, "Americans on St. Maarten Tell of Irma's Devastation, Lawlessness; 1,200 Evacuated," ABC 7 News, September 10, 2007; Dale Greer, "Airlift Operation Evacuates 1,000 U.S. Citizens from St. Maarten; 17 Kentucky Air Guardsmen Returning Home Today, 45 More Deploying to Support Irma Rescue Missions," 123rd Airlift Wing Public Affairs, September 11, 2007; Doug G. Ware, "Navy Moves Ships to Red Sea in Case US Needs to Move Americans Out of Sudan," *Stars and Stripes*, April 24, 2023; James G. Foggo, "Evacuating Sudan: An Amphibious Gap and Missed Opportunity," *Defense News*, May 3, 2023; Laura Gozzi and Alys Davies, "Sudan Fighting: Diplomats and Foreign Nationals Evacuated," BBC News, April 24, 2023; Arwa Ibrahim and Dalia Hatuqa, "Sudan Updates: Warring Sides Agree to Extending Truce," Al Jazeera, April 30, 2023; Mallory Shelbourne, "U.S. Navy Sends Nontraditional Ships to Support Sudan Evacuation," USNI News, May 1, 2023.

NOTE: USNS = U.S. Naval Ship; USS = U.S. Ship.

Box 3.2. Overseas Humanitarian, Disaster, and Civic Aid

The Overseas Humanitarian, Disaster, and Civic Aid (OHDACA) program is DoD's main humanitarian tool. It is an appropriation that "supports DoD and the Combatant Commanders' theater strategies to build partner nation capacity and expand and strengthen alliances and partnerships while advancing DoD access, influence and visibility." OHDACA covers several areas and lines of effort, including humanitarian assistance, humanitarian mine assistance, and foreign disaster relief. OHDACA falls under Title 10 of the U.S. Code, specifically sections 2561, 2557, 407, 404, and 402. OHDACA funds activities that benefit civilian populations in partner countries, as opposed to partner militaries. Therefore, OHDACA is "a military-to-civilian program to complement military-to-military security cooperation."

Under the humanitarian assistance line of effort, DoD "conducts collaborative engagements with partner nations to reduce endemic conditions such as human suffering, disease, hunger, and privation in regions where humanitarian needs pose challenges to stability, prosperity, and respect for universal human values." The humanitarian mine assistance program focuses on "landmines, explosive remnants of war, and unexploded ordinance." OHDACA also provides military component support to the interagency, specifically DoS and USAID, during foreign disaster relief missions.

The Office of the Secretary of Defense (OSD) oversees OHDACA, and the Defense Security Cooperation Agency implements the programs in tandem with the combatant commands. OHDACA funding from FY 2019 through FY 2022 is available on the DoD Open Government website.

SOURCES: OSD, *Fiscal Year (FY) 2022 President's Budget: Justification for Security Cooperation Program and Activity Funding*, May 2021, p. 43; DoD Open Government, "Security Cooperation," webpage, U.S. Department of Defense, undated.

MILITARY OPERATIONS AND THEIR COSTS

Endnotes

1. Chandler et al., 2023; Toukan et al., 2023.
2. Chandler et al., 2023; Miro et al., 2023; Palik et al., 2020, p. 8.
3. For more information about the dataset, including the criteria for inclusion, and RAND researcher analysis, see Jennifer Kavanagh, Bryan Frederick, Matthew Povlock, Stacie L. Pettyjohn, Angela O'Mahony, Stephen Watts, Nathan Chandler, John Speed Meyers, and Eugeniu Han, *The Past, Present, and Future of U.S. Ground Interventions: Identifying Trends, Characteristics, and Signposts*, RAND Corporation, RR-1831-A, 2017.
4. Kavanagh et al., 2017, p. 187.
5. *Armed conflict* is defined in this report as "[i]nterventions involving traditional military operations and fighting, characterized by large formations of organized military forces on both sides" (Kavanagh et al., 2017, p. 195). Note that these findings are drawn from a RAND study that primarily considers ground interventions (i.e., not naval or air interventions) because the majority of U.S. interventions in RUMID (about 75 percent) involve ground forces. Interventions that did not include ground components tended to be disproportionately deterrence interventions, often consisting of the time-limited deployment of a carrier strike group or similar naval capability intended to signal a desire for further escalation or outside intervention in an ongoing conflict or crisis.
6. Kavanagh et al., 2017, p. 30.
7. For example, RUMID shows that only two HADR operations occurred in the CENTCOM AOR from 1898 to 2017 as a result of the force size threshold. See Bryan Frederick, Jennifer Kavanagh, Stephanie Pezard, Alexandra Stark, Nathan Chandler, James Hoobler, and Jooeun Kim, *Assessing Trade-Offs in U.S. Military Intervention Decisions: Whether, When, and with What Size Force to Intervene*, RAND Corporation, RR-4293-A, 2021.
8. Miro et al., 2023.
9. Kavanagh et al., 2017, p. 41.
10. Of all the interventions included in the dataset, more than 70 percent of combat interventions involved at least 20,000 troops, while less than half of all COIN and stability operations involved 20,000 troops (Kavanagh et al., 2017, p. 23).
11. Toukan et al., 2023.
12. Kavanagh et al., 2017, p. 142.
13. Chandler et al., 2023.
14. Stefanie Mavrakou, Emelie Chace-Donahue, Robin Oluanaigh, and Meghan Conroy, "The Climate Change–Terrorism Nexus: A Critical Literature Review," *Terrorism and Political Violence*, Vol. 34, No. 5, 2022.
15. Jon Barnett and W. Neil Adger, "Climate Change, Human Security and Violent Conflict," *Political Geography*, Vol. 26, No. 6, August 2007.

16 Ashton Kingdon and Briony Gray, "The Class Conflict Rises When You Turn Up the Heat: An Interdisciplinary Examination of the Relationship Between Climate Change and Left-Wing Terrorist Recruitment," *Terrorism and Political Violence*, Vol. 34, No. 5, 2022.

17 Tobias von Lossow, *Water as Weapon: IS on the Euphrates and Tigris*, German Institute for International and Security Affairs, January 2016.

18 The *2018 National Defense Strategy* stresses that "[i]nter-state strategic competition, not terrorism, is now the primary concern in U.S. national security" (U.S. Department of Defense, *Summary of the 2018 National Defense Strategy of the United States of America*, 2018, p. 1). The same emphasis was carried over into the *2022 National Defense Strategy*.

19 We excluded personnel and fixed capital costs (such as weapon system acquisition and installation construction costs) with the assumption that the personnel and capital required to carry out targeted drone campaigns were already employed or acquired by DoD at the start of each operation. This is a conservative assumption, and we could therefore be underaccounting for the total costs of these campaigns.

20 For example, the model developed by RAND researchers suggests that, on average, a context with 10,000 refugees has a 4.1 percent likelihood of U.S. intervention, a context with 100,000 refugees has a 5.7 percent chance of intervention, and a context with 1 million refugees has a 7.7 percent chance of intervention (Kavanagh et al., 2017, p. 153).

21 Kavanagh et al., 2017, pp. 151–152.

22 Kavanagh et al., 2017, p. 156.

23 Heather Peterson, "U.N. Peacekeeping Is a Good Deal for the U.S.," *RAND Blog*, April 2, 2017.

24 Thomas Melito, *UN Peacekeeping: Cost Estimate for Hypothetical U.S. Operation Exceeds Actual Costs for Comparable UN Operation*, Government Accountability Office, GAO-18-243, February 2018; Joseph Christoff, *Peacekeeping: Cost Comparison of Actual UN and Hypothetical U.S. Operations in Haiti*, Government Accountability Office, GAO-06-331, February 2006.

25 This dataset includes U.S. HADR interventions from 1970 through 2003 drawn from Cobble, Gaffney, and Gorenburg, 2005, and U.S. HADR interventions from 2003 through 2023 from a search of USAID reporting (USAID is the lead U.S. government agency on HADR).

26 A few examples of short interventions: flood relief and insecticide spraying in Pakistan in 1973 (one month), Yemen earthquake relief in 1981 (one day), the provision of firefighting equipment for oil field fires in Uzbekistan in 1992 (one day), and Iran earthquake relief in 2004 (about 16 days).

27 For example, Yemen earthquake relief in 1981 involved six C-141s, and Iran earthquake relief in 2004 involved seven C-130s.

28 Ellison and Sikorsky, 2023.

29 Jess T. Ford, *State Department: Evacuation Planning and Preparations for Overseas Posts Can Be Improved*, Government Accountability Office, GAO-08-23, October 2007b.

30 Ford, 2007b.

31 Ford, 2007b.

CHAPTER 4

CONCLUSION

THROUGHOUT THIS REPORT, we highlighted how and when CENTCOM could be called on to either reduce the risk of climate-related conflict or respond to that conflict. Through this analysis, we identified six key findings to support the command as it faces the effects of climate change and the impact such change could have on regional security.

Key Findings

First, over the coming decades, climate stressors will become more intense and more frequent in the CENTCOM AOR. This will likely contribute to CENTCOM's broader shift from a warfighting-focused command to a command that will have to reprioritize and balance how it responds to and conducts both traditional and nontraditional security concerns. Through partnerships and innovation, CENTCOM can support regional efforts to mitigate the risk of climate-related conflict along with supporting foreign partners' endeavors to become climate resilient.

Second, CENTCOM will likely play a supporting role to interagency partners in reducing the risk of climate-related conflict because the causal pathways from climate hazards to conflict revolve around political and economic concerns that will need to be addressed if future conflicts are to be averted. However, military-led OAIs provide some niche tools for interrupting the progression from climate hazards to conflict and could decrease the severity of conflicts by improving U.S. and partner response capabilities. The OAIs identified in this report focused on nontraditional security activities; however, there are opportunities in which traditional security cooperation activities could have a secondary effect on climate-related adaptation activities.

Third, in addition to mitigating conflict risk, CENTCOM has an opportunity to develop partner resilience to climate change for the ancillary benefit of strengthening bonds within the CENTCOM coalition. This is an important consideration for CENTCOM because of the United States' interest in broadening cooperation between long-standing CENTCOM partners and its newest coalition member, Israel, which only joined CENTCOM in 2021 after overcoming long-standing sensitivities from Arab states. CENTCOM can continue to leverage long-standing multilateral forums as well, while considering the usefulness of establishing new consultative bodies that advance technological sharing as it relates to climate adaptation and resiliency.

Fourth, in the coming decades, there is likely to be an increased demand on CENTCOM to assist with HADR in the theater. CENTCOM can help partners build response capabilities, although CENTCOM should also be prepared to support these operations directly when required. Pakistan has historically been a major focus of these operations and will likely continue to be, based on climate projections. Moreover, understanding the climate-to-conflict pathways, coupled with climate and conflict projections for the region, will help intelligence professionals and military planners to anticipate which parts of the AOR could see an increase in HADR operations.

Fifth, if climate stress manifests in more conflicts in the CENTCOM AOR and U.S. policymakers define U.S. interests in ways that lead Washington to intervene, then stabilization operations would likely drive the highest costs for the operation types we considered in our costing analysis. DoD costs for the NATO-led stabilization operations we considered in our analysis number in the tens of billions of dollars per operation. On the basis of our climate projections, stabilization forces will have to contend with more extreme heat and less available water, among other climate hazards, which will likely stress regional logistics capabilities.

Finally, HADR, counterterrorism operations, NEOs, and planned security cooperation (i.e., from the base budget rather than overseas contingency operations funding) impose fewer costs on DoD funding. In the sample of cases we considered in our costing analysis, the DoD-borne costs of these operations never exceeded $1 billion, and the annual costs of the counterterrorism missions and the DoD-borne costs of HADR operations in our sample were typically just tens of millions of dollars. Although such operations occur at a frequency that does not stress current budgets, the financial calculus could change if climate hazards lead to requirements for increased HADR operations or NEOs.

Key Recommendations

On the basis of these findings, we make the following recommendations. These recommendations complement the OAIs that were presented in Table 2.1 and, in some instances, expand on them.

First, with national security priorities and budgets shifting toward other GCCs, CENTCOM planners will be required to sharpen their prioritization of competing demands for resources. Budget reductions include decreasing resources for security cooperation and the elimination of overseas contingency operations funding, which has been a primary source of the CENTCOM security cooperation budget (a detailed analysis can be found in the appendix). When considering security cooperation investments with partners, we recommend prioritizing nontraditional security cooperation activities because they address emerging threats, draw on the United States' strength in technological adaptation, and can be leveraged to deepen cooperation within the CENTCOM coalition. OAIs related to climate preparedness and resilience that support off-ramp actions away from potential climate-related conflict support CENTCOM's focus areas of partnerships and innovation.

Second, CENTCOM planning and intelligence staff should incorporate climate intelligence and analysis, including time frame–appropriate climate projections and other related analyses, when developing or updating theater campaign plans, operational plans (OPLANs), and CONPLANs. This includes updating CONPLANs for HADR operations and NEOs in light of future climate projections and their effect on the physical and security environments. Moreover, by including climate analysis in all OPLANs and CONPLANs, the command will have a better understanding of climate constraints on forces and equipment, along with changes to the physical environment. Intelligence officers and analysts can be trained to identify the climate-related developments and signposts that could lead to humanitarian disasters or contribute to political instability. Additionally, intelligence analysis can support planners by identifying specific hot spots that could be emerging priorities for OAIs within the planning horizon for OPLANs and CONPLANs. Identifying more near-term climate stressors and the locations of those stressors within the AOR, along with possible OAIs as off-ramps or hedges, could aid in command efforts to mitigate the risk of climate-related conflict.

Third, CENTCOM should request the expansion of the SPP within the AOR, particularly with National Guard units that have experience in climate resiliency and adaptation. Critically, the development of an SPP with Pakistan, in which the National Guard partner has experience with disaster response related to flooding, should be prioritized considering the frequency of HADR operations in that country over the past few decades.

Finally, CENTCOM would benefit from greater climate literacy among its headquarters (HQ) staff and among forward-deployed personnel, such as security cooperation officers and defense attachés based in the region. Education and training should occur at multiple levels, including at joint professional military education institutions, the Defense Security Cooperation University, and CENTCOM HQ. This education and training would support key leader engagements and critical discussions with partner nations about the projected impacts of more intense and more frequent climate hazards, facilitating discussions about enabling climate resiliency in the AOR. Additionally, CENTCOM HQ personnel would be able to better incorporate climate analysis into intelligence assessments, military construction, and military plans as a result of this education and training.

Areas for Further Research

In addition to our findings and recommendations, our research has revealed several areas for future research. These proposed research areas would assist CENTCOM, and potentially additional DoD audiences, with understanding the effects of climate change on the security environment and how military responses might have to adapt over the next few decades.

First, the share of CENTCOM's attention and resources that will be devoted to climate-related issues will also depend on the evolution of traditional security threats in the theater. The future of Iran as a security threat to U.S. partners and U.S. forces forward deployed and the future of (in)stability within Iran are critical elements of traditional security threats in the theater. Similarly, the future of U.S. relationships with key CENTCOM partners and their relationships with other extra-regional actors (e.g., China and Russia) will have great bearing on the security environment. Analyzing these drivers of military requirements while incorporating climate intelligence and analysis is required to better understand the future balance in the theater between traditional and nontraditional security issues.

Second, our research has not addressed the potential effect of successful or failed green transitions on regional security, and this issue could have the greatest impact on how climate change ultimately affects the CENTCOM security environment. Because the governance model of key U.S partners is based on a rentier system linked to revenue from hydrocarbons, these states' ability to adapt to shifts in global hydrocarbon demand and eventually navigate green transitions and associated costs warrants further research as a major factor in regional stability.

Third, increasing the access of CENTCOM personnel to climate literacy education and training depends on the composition of the CENTCOM workforce. For example, field grade officers at CENTCOM have access to professional military education that civilians and contractors do not, but civilians and contractors might have easier access to reimbursable education from civilian educational institutions. On the other hand, uniformed personnel would be expected to cycle through the command more quickly than other elements of the workforce. More research is required that includes a detailed analysis of the CENTCOM workforce to provide precise recommendations on how CENTCOM can best develop greater climate literacy within the command, which will aid analysis of changing climate effects on the regional security environment. One potential option is the establishment of a cell or center at CENTCOM HQ that focuses on climate-related security issues to provide enduring expertise. This cell or center can also serve as an in-house education center for uniformed personnel on their regular tours.

Fourth, our research only briefly touched on the impacts of climate hazards, without a full analysis of how those hazards could affect nontraditional security elements. The effects could include migration movements within the AOR and out of the region, public health trends, and resource management, including impacts on domestic and imported agricultural production. Climate change will affect each of these differently, which will have corresponding consequences on the regional security environment and CENTCOM's ability to respond.

Fifth, our costing analysis was not scoped to include the costs of making existing military installations more resilient to climate impacts via adaptation or potentially relocating installations to new facilities to minimize vulnerability to extreme weather events and extreme heat. That could be a major driver of DoD military construction costs in the CENTCOM AOR and merits further research.

Finally, our design of the costing analysis in this study was intended—with one exception—to extrapolate from the historic costs by operation type rather than enumerate all possible costs given assumptions about the requirement. The latter approach could reveal much higher costs than our design if climate stresses lead to much larger impacts that shift the scale of HADR from historic patterns. Historically, the smaller HADR and NEO responses have not stressed readiness levels; however, the frequency of such operations could increase, which could cause additional pressures and costs on forces.

APPENDIX

DETAILED ROUGH ORDER OF MAGNITUDE COSTING

THIS APPENDIX PROVIDES more details for a selection of the tables provided in the costing section on the following interventions and activities: counterterrorism, stabilization, HADR, and NEOs. We provide empirical costing data for select past operations, which establish a useful baseline of costs (and, in some instances, force requirements) for how the United States has used military forces. This analysis also provides a point of departure and comparison for considering the costs of future military engagements in the AOR over the next decades.

Additionally, this appendix starts with a discussion regarding current security cooperation funding levels. This was not included in the Chapter 2 discussion on building partnerships because most of the OAIs that were recommended are not currently incorporated in security cooperation budgets. However, understanding security cooperation funding trends—and in comparison with other GCCs—is critical for CENTCOM as it faces a security environment that is more heavily influenced by climate hazards than in previous decades.

Security Cooperation Costing

DoD defines *security cooperation* as

> [a]ll Department of Defense interactions with foreign security establishments to build security relationships that promote specific United States security interests, develop allied and partner nation military and security capabilities for self-defense and multinational operations, and provide United States forces with peacetime and contingency access to allied and partner nations.[1]

Activities include, but are not limited to, military engagements with foreign defense and security establishments (including those primarily performing disaster or emergency response functions), DoD-administered security assistance programs (which occur in consultation with DoS), combined exercises, international armaments cooperation, and information-sharing and collaboration.[2] The FY 2017 National Defense Authorization Act required the President to submit a formal, consolidated budget request for DoD's security cooperation efforts beginning with the FY 2019 budget.[3] In Table A.1, we present the President's budget request to Congress for security cooperation programs and activities by GCC from FY 2019 to FY 2022.[4] These costs include a variety of security cooperation activities, such as training with foreign forces, capacity-building, and military personnel exchange programs. The costs exclude cooperation activities associated with classified programs and drug interdiction operations. In line with the budget requests, we break out security cooperation by base funds and contingency funds associated with combat or direct combat support.

As shown in Table A.1, contingency funding requests for security cooperation, which were as high as $7.8 billion from FY 2019 to FY 2022 for CENTCOM, far exceeded base funding requests, which ranged from $108 million to $429 million per command during this period. The overwhelming majority of CENTCOM's contingency funds during this period supported operations in Afghanistan and Iraq. As of FY 2022, overseas contingency operations funding was eliminated and incorporated into the main DoD budget.[5] Through our off-ramp and OAI analyses, we determined that nontraditional security cooperation activities are a likely area where CENTCOM can engage with partners to reduce the risk associated with climate-related conflict. Box A.1 details significant DoD security cooperation initiatives.

Counterterrorism: Detailed Table

To complement the overall analysis provided in Table 3.1, Table A.2 provides the detailed accounting and analysis of costing related to counterterrorism drone operations. The costing model takes as its unit of analysis a combat air patrol (CAP) unit, also known as a *drone orbit*. A CAP typically consists of four MQ-1 Predator or MQ-9 Reaper drones, one of which is kept flying at any given moment to maintain a 24-hour, seven days per week view of the target area. We applied this model to four U.S. counterterrorism campaigns characterized by targeted drone attacks—Somalia (2011–2018), Libya (2012–2019), Yemen (2009–2013), and Pakistan (2004–2018)—by extrapolating the variety of CAPs deployed

Table A.1. Estimated Geographic Combatant Command Security Cooperation Costs, Fiscal Year 2019 to Fiscal Year 2022 (in 2022 dollars)

		FY 2019	FY 2020	FY 2021	FY 2022
CENTCOM	Base	$208,181,397	$157,416,527	$157,010,588	$108,138,833
	Contingency	$7,791,180,273	$7,662,350,790	$6,036,423,839	$4,655,332,000
SOUTHCOM	Base	$230,957,951	$312,230,494	$197,403,192	$240,054,833
	Contingency	$372,113,005	$0	$111,877,266	$0
EUCOM	Base	$230,872,830	$173,044,041	$199,139,726	$182,717,833
	Contingency	$596,115,311	$549,652,143	$379,200,920	$410,029,000
AFRICOM	Base	$241,247,497	$185,786,782	$177,918,505	$156,739,833
	Contingency	$372,113,005	$185,150,511	$111,877,266	$100,379,000
INDOPACOM	Base	$428,953,590	$172,740,090	$214,846,595	$302,519,833
	Contingency	$372,113,005	$0	$111,877,266	$145,988,000
NORTHCOM	Base	$209,427,970	$428,126,546	$177,036,337	$128,667,833
	Contingency	$372,113,005	$198,397,258	$111,877,266	$0

SOURCES: OSD, 2018; OSD, 2019; OSD, 2020; OSD 2021.

NOTE: The amounts reported in this table are requested and not enacted funds because the budget requests only include three rather than four years of enacted data. For FY 2021, the requested funds for security cooperation amounted to about $7.6 billion, while the enacted amount was $6.7 billion. For programs and activities not specific to a single GCC, we divided the requested funds by the total number of GCCs to approximate the amount of funds allocated to each command.

Base funding includes the following programs and activities: African Partnership Flight, African Partnership Station, military-to-military engagements, Pacific Partnership, SPP (National Guard), joint combined exchange training, training with friendly foreign countries, support to conduct of operations, International Security Cooperation Programs account, Aviation Leadership Program, Inter-American Air Forces Academy, Inter-European Air Forces Academy, regional centers for security studies, Regional Defense Fellowship Program, service academy international engagement, Western Hemisphere Institute for Security Cooperation, Defense Institute for International Legal Studies, Institute for Security Governance, security cooperation strategic evaluations, security cooperation program management, security cooperation workforce development, security cooperation data management, Defense Institute of Security Cooperation Studies, and cooperative threat reduction. Most, but not all, of these accounts were included in the budget justifications for FY 2019 to FY 2022.

Contingency funding includes the following programs and activities: border security, coalition support program, lift and sustain, International Security Cooperation Programs account, Afghanistan Security Forces Fund, Counter-Islamic State of Iraq and Syria Train and Equip Fund, and the Ukraine Security Assistance Initiative.

from the average annual number of U.S. air strikes for each of these campaigns. For campaigns with a similar number of average air strikes as the Somalia campaign, which we used as a baseline, we assumed the deployment of two to four CAPs. For campaigns with a larger number of average air strikes, we assumed the deployment of a higher range of CAPs. Average air strikes were generally collected from two primary sources: New America and the Bureau of Investigative Journalism.[6] The average number of air strikes was also used to estimate munition costs. Finally, we used New America data on the number of civilian casualties associated with U.S. air strikes to calculate average condolence and solatia payment costs for each campaign.[7]

As shown in Table A.2, total operational costs are primarily driven by the estimated range of CAPs deployed and the length of the conflict. On the basis of our sample, we estimate the range for total operational costs of a targeted drone campaign for one year to be between approximately $10 million (Somalia and Yemen) and $30 million (Libya and Pakistan).

Stabilization: Detailed Tables

Tables A.3 and A.4 provide the specific cost data that are represented in Figure 3.4. These tables also offer descriptions of the specific stabilization operations that occurred and a summary of the purpose and mission of those operations. Table A.3 additionally details information on force requirements and activities that those forces conducted while supporting the stabilization operations. These tables provide planners with an overview of OAIs that CENTCOM might be called on to support more frequently in the future. As noted in Chapter 3, costing information for operations in Afghanistan and Iraq are not included because those missions evolved over the course of multiple decades.

Humanitarian Assistance and Disaster Response: Detailed Table

Table A.5 furnishes the specific cost data that is represented in Figures 3.4 and 3.5. The table additionally describes specific HADR operations, which include a summary of the purpose and mission, along with examples of the assistance DoD provided. This table gives insights to planners on the variety of HADR operations that CENTCOM might need to prepare for.

Box A.1. Significant Security Cooperation Initiatives

Significant security cooperation initiatives (SSCIs) were created by OSD after the 2017 National Defense Authorization Act reforms. These reforms aimed to ensure that DoD security cooperation efforts are aligned with the National Security Strategy and the National Defense Strategy and that those efforts are making a good return on investment. SSCIs are defined by OSD as

> [t]he series of activities, projects, and programs planned as a unified, multi-year effort to achieve a single desired outcome or set of related outcomes. Such initiatives are generally planned by the geographic Combatant Commands and involve the application of multiple security cooperation tools over multiple years to realize a country- or region-specific objective or functional objective as articulated in the country-specific security cooperation sections of a theater campaign plan.[a]

SSCIs are used to prioritize annual spending for the International Security Cooperation Programs (ISCP) account, which funds "activities aimed at building partner capacity to address shared national security challenges and operate in tandem with or in lieu of U.S. forces."[b] ISCP includes train-and-equip programs (333) and institutional capacity-building (332), which are aimed at increasing the absorptive capacity and sustainment capabilities of partners. ISCP also includes the Indo-Pacific Maritime Security Initiative. SSCIs funded by ISCP go through a rigorous review to ensure that security cooperation efforts "are prioritized across a variety of factors, including strategic alignment, program feasibility, and DoD component prioritization."[c] All SSCIs require an assessment, monitoring, and evaluation component, including initiative design documents, performance monitoring plans, and, increasingly, strategic evaluations. In Table A.1, these funds are tallied under contingency funding.

OSD oversees the SSCI process, and the Defense Security Cooperation Agency implements the programs in tandem with the combatant commands. ISCP funding from FY 2019 through FY 2022 is available on the DoD Open Government website.[d]

[a] Department of Defense Instruction 5132.14, *Assessment, Monitoring, and Evaluation Policy for the Security Cooperation Enterprise*, Office of the Under Secretary of Defense for Policy, January 13, 2017, p. 23.
[b] OSD, 2021, p. 25.
[c] OSD, 2021, p. 25.
[d] DoD Open Government, undated.

APPENDIX

Table A.2. Modeled Historic Costs of Selected Targeted Drone Campaigns, 2004 to 2019

		Somalia 2011–2018	Libya 2012–2019	Yemen 2009–2013	Pakistan 2004–2018
		• Two to four CAPs • 16.4 average annual air strikes • Four to five average annual civilian casualties	• Three to six CAPs • 27.4 average annual air strikes • One to two average annual civilian casualties	• Two to four CAPs • 21.2 average annual air strikes • 17–18 average annual civilian casualties	• Three to six CAPs • 28.1 average annual air strikes • 21–22 average annual civilian casualties
Mission flight costs[a]					
MQ-9 Reaper O&M, MQ-1 Predator O&M	$4.1–5.1 million annually per CAP	$8.2–20.4 million annually	$12.3–30.6 million annually	$8.2–20.4 million annually	$12.3–30.6 million annually
Munition costs[b]					
AGM-114 Hellfire II air-to-surface missiles	$89,000 per munition (approximately 95% of all munitions)	$1.4 million annually	$2.3 million annually	$1.8 million annually	$2.4 million annually
500-pound joint direct attack munition bomb	$29,000 per munition (approximately 5% of all munitions)	$29,000 annually	$58,000 annually	$29,000 annually	$58,000 annually
Civilian casualty costs					
Condolence and solatia payments[c]	$5,000 per casualty	$20,000–25,000 annually	$5,000–10,000 annually	$85,000–90,000 annually	$105,000–110,000 annually
Total operational costs (annual)		$9.7–21.9 million	$14.7–33 million	$10.1–22.3 million	$14.9–33.2 million
Total operational costs (all years)		$77.7–175 million	$117.4–263.9 million	$50.5–111.5 million	$223–497.6 million

SOURCES: Bergen, Sterman, and Salyk-Virk, 2021; Serle and Purkiss, 2017; Turse, Moltke, and Speri 2018; McAndrew, 2018; Anderson et al., 2018.

NOTE: The time frame of each campaign is based primarily on the years for which data on air strikes are available. For Yemen, we focused on the period between 2009 and 2013, after which U.S. involvement shifted from a counterterrorism campaign to an intervention in a civil war. Average annual air strikes are derived from data from New America and the Bureau of Investigative Journalism, which publish data on U.S. air strikes for various countries. As the number of air strikes from these two sources sometimes varies, we took the average to estimate annual air strikes. The Bureau of Investigative Journalism does not collect data for Libya. Accordingly, we used numbers in an article published by The Intercept as our second source. New America does not disaggregate drone strikes from other types of air strikes. Where possible, we limited our sample to strikes where no more than three munitions were released, with the assumption that these types of strikes are more likely to be associated with drones rather than fighter jets. For one country, we also reviewed a random sample of ten air strikes and determined that the strikes in our sample were all carried out by drones.

[a] We assume that, in total, a CAP unit must fly about 7,300–8,760 hours per year to maintain a continuous stare over a target area, based on Hasik and our calculation of estimated flight hours per CAP in which at least one drone is airborne and providing a continuous stare over a target area (James Hasik, "Affordably Unmanned: A Cost Comparison of the MQ-9 to the F-16 and A-10, and a Response to Winslow Wheeler's Criticism of the Drone," *James Hasik: Thinking on Innovation, Industry, and International Security*, blog, June 20, 2012). We use this annual total, along with Office of the Under Secretary of Defense Comptroller data on estimated hourly flight costs for Predators and Reapers ($557 and $577 per flight hour, respectively), to infer annual O&M mission flight costs for a single Reaper or Predator CAP unit of $4.1 million to $5.1 million per year.

[b] Reaper and Predator drones can be armed with 100-pound AGM-114 Hellfire II air-to-surface missiles, 500-pound joint direct attack munition bombs, or a combination of these two types of munitions. As the ratio of Hellfire missiles to 500-pound smart bombs released during air strike operations is not knowable in an unclassified setting, we use an estimate based on data on total Reaper and Predator combat operations conducted by the 432nd Wing at Creech Air Force Base from 2011 to 2016, which suggests that no more than 5 percent of weapons released during U.S. drone campaigns are 500-pound smart bombs.

[c] Condolence and solatia payments are payments made to compensate for the killing of civilians by U.S. forces during a conflict. In line with McNerney et al., who report that the average annual condolence payments made in Iraq between 2015 and 2019 ranged from $3,000 to $5,000, we use an estimate of $5,000 for each civilian death (McNerney et al., 2021).

Table A.3. NATO and U.S. Stabilization Examples, 1999 to 2003 (in 2022 dollars)

Case	Short Description	Examples of NATO Activities	Total NATO Costs	DoD Contributions
Kosovo (1999)	After the 1999 Kosovo War, NATO's KFOR contributed to stabilization policies broadly, including policing, justice and immigration systems, and transition to democratic governance.	47,000 NATO KFOR troops, including an initial 8,000 U.S. troops, were deployed to stabilize Kosovo, provide security to UNMIK, and provide safety and security to Kosovo's inhabitants in conjunction with the UN International Police. KFOR also supported the demobilization and disbanding of the KLA and helped reintegrate former KLA members into society through counseling, job training, and other interventions. Other activities included overseeing disarmament, clearing minefields, controlling borders, and protecting freedom of movement.	$19,338,376,151	$12,915,946,524
Bosnia (1995–2003)	After the Dayton Accords ended the war in Bosnia and Herzegovina, NATO was tasked with preventing the resumption of hostilities, maintaining peace and security, and contributing to national reconstruction by building up governance, policing, and military capacities. The NATO IFOR oversaw the period immediately after the Treaty of Dayton that ended the war, until the situation stabilized and IFOR was replaced by the smaller NATO SFOR.	60,000 NATO IFOR troops, about one-third of which were U.S. troops, were deployed in 1995 and 1996 to monitor and enforce a zone of separation along the inter-entity boundary line between the Bosnian Federation and the Republika Srpska (the two entities defined by the peace agreement). IFOR also oversaw the removal of weapons to approved cantonment sites, cleared minefields, and responded to any violent incidents. NATO SFOR was responsible for continuing to deter the resumption of hostilities, arresting individuals indicted for war crimes, assisting the return of refugees and displaced people, and supporting civilian implementation efforts, including the holding of postponed municipal elections in mid-1997.	$87,657,462,893	$18,934,011,985

SOURCES: NATO, "NATO Mission in Kosovo (KFOR)," webpage, undated; Sletzinger, 2003; Bureau of European and Eurasian Affairs, "NATO's Role in Bosnia and Herzegovina," fact sheet, U.S. Department of State, December 6, 2004; Micah Zenko, "What Does Libya Cost the United States?" *Politics, Power, and Preventive Action*, Council on Foreign Relations blog, August 11, 2011; Julie Kim, *Bosnia Implementation Force (IFOR) and Stabilization Force (SFOR): Activities of the 104th Congress,* Congressional Research Service, No. 96-723, January 6, 1997; White House, "Winning the War and the Peace in Kosovo," undated; Ministry of Defense, Netherlands, "Kosovo Force 1999–2000 (KFOR)," webpage, undated; Zyla, 2015.

NOTE: IFOR = Implementation Force; KFOR = Kosovo Force; KLA = Kosovo Liberation Army; SFOR = Stabilization Force; UNMIK = UN Interim Administration Mission in Kosovo. NATO and U.S. costs in Kosovo include an 11-week air campaign against Serbia.

Noncombatant Evacuation Operation: Detailed Table

Table A.6. provides expanded discussion to the information that is represented in Table 3.2. Unlike the other operations, we were unable to determine the costs of NEOs. Instead, we provide details on the forces and systems required to conduct the operation as a stand-in for a costing estimate. This table gives insights to planners on the variety of NEOs that CENTCOM might need to prepare for.

Table A.4. UN Stabilization (U.S. Cooperating) Examples, 1999 to 2012 (2022 dollars)

Case	Short Description	UN Costs	U.S. Approximate Share
UN Mission in Sierra Leone (UNAMSIL, 1999–2005)	UNAMSIL enforced the Lomé Peace Accords that followed the Sierra Leone Civil War with a sizable military presence. The mission was expanded in 2000 and 2001 to continue providing broad protection and assistance. Activities included supporting elections and disarmament, protecting transportation routes, and providing humanitarian assistance.	$4,908,817,922	$1,374,469,018
UN Transitional Administration in East Timor (UNTAET, 1999–2002)	UNTAET oversaw East Timor's independence process. The mission provided stability, order, and humanitarian aid and oversaw the formation of a new government.	$786,413,222	$220,189,105
UN Organization Mission in the Democratic Republic of the Congo (MONUC, 1999–2010)	MONUC helped stabilize the Democratic Republic of the Congo in the wake of its civil war by overseeing the cessation of hostilities. The mission was tasked with overseeing the Lusaka Agreement ceasefire and accomplished this while providing humanitarian assistance, freeing prisoners, clearing mines, and working toward broad disarmament.	$14,398,883,034	$4,031,687,250
UN Mission in Ethiopia and Eritrea (UNMEE, 2000–2008)	UNMEE defended a demilitarized region near the border between Eritrea and Ethiopia, where a border conflict had been taking place. The mission supported demilitarization and humanitarian aid but ultimately withdrew without claiming success.	$2,132,741,112	$597,167,511
UN Mission of Support in East Timor (UNMISET, 2002–2005)	UNMISET retained a presence in East Timor after UNTAET to keep state functions operational during the turnover to the new East Timor government.	$878,265,897	$245,915,364
UN Mission in Côte d'Ivoire (MINUCI, 2003–2004)	MINUCI implemented a peace agreement that addressed the First Ivorian Civil War, including establishing a zone of confidence between the factions. This mission would lead to the longer UN Operation in Côte d'Ivoire, which ended successfully.	$45,564,437.49	$12,754,995
UN Operation in Burundi (ONUB, 2004–2006)	ONUB disarmed militants and protected civilians in response to concerns regarding the rise of armed groups and their ability to inflict violence on the population of Burundi. The mission also protected the movement of refugees and oversaw the upkeep of state processes, such as elections.	$1,009,016,979	$282,518,804
UN Mission in Sudan (UNMIS, 2005–2011)	UNMIS implemented the peace agreement signed between Sudan and the Sudan People's Liberation Movement, which lead to the creation of South Sudan. The mission oversaw the ceasefire and worked to rebuild the rule of law. The mission also assisted in the protection of human rights and the response to internal and external displacement.	$8,315,370,735	$2,328,303,806
UN Integrated Mission in Timor-Leste (UNMIT, 2006–2012)	UNMIT ensured the stability and functionality of East Timor (an expansion of UNTAET and UNMISET). This included monitoring human rights and capacity-building, especially in the justice system.	$1,883,360,290	$527,340,881
UN Mission in the Central African Republic and Chad (MINURCAT, 2007–2010)	MINURCAT responded to instability in refugee-heavy areas of Chad and in the neighboring regions of Sudan and the Central African Republic by protecting civilians, providing humanitarian aid, and facilitating refugee return. The mission ended at the request of the government of Chad.	$1,891,528,221	$529,627,902
UN Supervision Mission in Syria (UNSMIS, 2012)	UNSMIS oversaw a ceasefire in the Syrian civil war. The ceasefire ended before the oversight mandate, and the mission was suspended. UNSMIS was unable to accomplish its goal, and the mission ended unsuccessfully.	$21,156,598	$5,918,810

SOURCES: UN Integrated Mission in Timor-Leste, undated; UN Mission in Côte d'Ivoire, 2004; UN Mission in Ethiopia and Eritrea, 2009; UN Mission in Sierra Leone, 2005; UN Mission in the Central African Republic and Chad, undated; UN Mission in the Sudan, undated; UN Mission of Support in East Timor, 2005; UN Operation in Burundi, 2007; UN Organization Mission in the Democratic Republic of the Congo, undated; UN Transitional Administration in East Timor, 2002; UN Peacekeeping, "Promoting Human Rights," webpage, undated.

Table A.5. Examples of U.S. Humanitarian Assistance and Disaster Response Operations and Associated Department of Defense and Total U.S. Government Costs, Fiscal Year 1998 to Fiscal Year 2022 (in 2022 dollars)

Case	Short Description	Examples of DoD Assistance	DoD costs	Total USG Costs
Kosovo crisis (1998)	Following clashes between Serbian police forces and members of the Kosovo Liberation Army (KLA), Serbian police raided villages in Kosovo's Drenica region, resulting in thousands of Kosovar Albanians being displaced from their homes.	Construction of a 20,000-person refugee site in Fier, Albania, and transport and delivery of relief commodities, including at least 1 million HDRs.	$7,214,158	$138,331,224
East Timor civil conflict (1999)	Following a vote for independence from Indonesia, pro-integrationist militias in East Timor rampaged and plundered through large areas of East Timor. Hundreds of civilians were killed in the ensuing violence, and 450,000 East Timorese were displaced from their homes.	Donation of 300,000 HDRs to the World Food Program (WFP) for airdrop to isolated populations; transport of Portuguese relief commodities to Australia; refurbishment and transport of 20 trucks to East Timor for WFP's use; and the provision of air support to WFP for the delivery of food and non-food relief supplies.	$77,471,821	$842,192,366
Afghanistan —complex emergency (2001)	Two decades of war in Afghanistan resulted in an extended humanitarian crisis that was further exacerbated by a four-year regional drought. The drought destroyed local agriculture and livestock and forced affected populations to leave their homes in search of food and water.	Delivery of about 2.5 million HDRs.	$77,562,817	$1,190,086,549
Lebanon —complex emergency (2006)	Clashes between the Israel Defense Forces and Hezbollah resulted in the death of more than 300 people, injury of 800 people, and displacement of up to 500,000 Lebanese people.	Performance of unexploded ordnance clearance and demining activities.	$6,990,944	$152,885,082
Sri Lanka —complex emergency (2006)	Conflict intensification between the Sri Lankan government and the Liberation Tigers of Tamil Eelam separatist group that resulted in the displacement of an estimated 200,000 people, primarily in northern and eastern Sri Lanka.	Provision of medical supplies and equipment to primary health care centers and hospitals serving internally displaced persons.	$3,958,857	$371,365,066
Georgia—complex emergency (2008)	Armed conflict involving the Georgian government, Russian forces, and South Ossetian separatist forces that resulted in the displacement of about 200,000 people.	Delivery of relief commodities from DoS and DoD warehouses in Germany and USAID stockpiles in Italy.	$2,093,224	$76,267,285
Pakistan —complex emergency (2008)	Fighting between the Pakistani government and militant groups in Pakistan's North-West Frontier Province resulted in the displacement of more than 290,000 people across the province's nine districts.	Provision of meals, air-conditioned tents, and generators.	$22,424,039	$74,274,384
Influenza A (H1N1) (2009)	Sustained transmission of the influenza A (H1N1) virus in multiple Central American countries.	Transport of 30,600 personal protection equipment (PPE) kits from DoD stockpiles in Charleston, South Carolina, to Belize, El Salvador, Guatemala, Haiti, Honduras, and Nicaragua.	$295,603	$21,737,071

Table A.5—Continued

Case	Short Description	Examples of DoD Assistance	DoD costs	Total USG Costs
Pakistan flooding (2010)	Abnormally intense monsoon rains in Pakistan resulted in massive flooding. One-fifth of Pakistan's territory was flooded, and one in eight Pakistanis was directly affected.	Transport of more than 11,000 metric tons of relief supplies, evacuation of more than 26,000 people stranded by flooding, donation of 450,000 HDRs, and the provision of eight prefabricated bridges.	$70,253,820	$778,400,504
West Africa —Ebola outbreak (2014)	Outbreak of the Ebola virus disease, primarily in Sierra Leone, Guinea, and Liberia, which ended in 2016 following 28,600 cases and 11,325 deaths.	Design and construction of Ebola treatment units, medical laboratories, and other facilities; transport of medical supplies and personnel; provision of medical training to health care workers.	$766,432,638	$3,329,093,028
Iraq—complex emergency (2014)	Clashes between the Islamic State of Iraq and the Levant and its allied militias and the Iraqi government and Kurdish Regional Government throughout northern and central Iraq contributed to the displacement of a large number of people, many from Yazidi and other minority groups.	Provision of seasonally appropriate items, such as winter clothing for approximately 550,000 children, to help affected populations prepare for the winter months.	$92,787,096	$3,300,125,615
Pakistan flooding (2022)	Flooding and landslides caused by heavy rains and glacial lake outbursts resulted in the death of more than 1,700 people and affected 33 million people.	Transport of 1.4 million pounds of relief commodities from USAID's Dubai warehouse to Pakistan.	$1,985,619	$98,434,081

SOURCES: USAID, 1999a; USAID, 1999b; USAID, 2000a; USAID, 2000b; USAID, 2003; USAID, 2006; USAID, 2007a; USAID, 2007b; USAID, 2009a; USAID, 2009b; USAID, 2009c; USAID, 2009d; USAID, 2009e; USAID, 2011; USAID, 2020; USAID, 2022; USAID, 2023; Moroney et al., 2013; Joint and Coalition Operational Analysis, "Operation UNITED ASSISTANCE: The DoD Response to Ebola in West Africa," January 6, 2016.

NOTE: HDR = humanitarian daily ration.

Table A.6. Examples of Noncombatant Evacuation Operation Requirements, 2003 to 2023

Case	Short Description	Forces and Systems Employed
Operation Shining Express (Liberia, 2003)	Thirty-five combat-equipped U.S. military personnel were deployed to Monrovia, Liberia, to augment U.S. embassy security forces and aid in the possible evacuation of U.S. citizens. Fixed-wing and rotary aircraft and their crews were also prepositioned in Dakar, Senegal, and Freetown, Sierra Leone, to assist with a larger, more-complex evacuation operation if conditions worsened. USS *Kearsarge*, along with its 3,000 sailors and marines, was redirected to the West African coast in support of a potential large-scale evacuation.	• USS *Kearsarge* (amphibious assault ship) • Three HH-60 Pave Hawk helicopters and their crews from the 85th Group at Naval Air Station Keflavik, Iceland • One MC-130P Combat Shadow plane from the 352nd Special Operations Group from Royal Air Force Mildenhall, England
Lebanon (2006)	In coordination with DoS, DoD evacuated close to 15,000 U.S. citizens from a war zone in less than a month. This was the largest U.S. government NEO since the 1979 evacuation from Iran. The operation involved transporting or escorting thousands of evacuees via contracted commercial ships, U.S. naval ships from the 24th Marine Expeditionary Unit, and helicopters, from Beirut to Cyprus and Incirlik Airbase in Turkey. DoD was also instrumental in arranging flights for Americans from safe haven locations in Cyprus and Turkey to the United States. At least 2,200 marines and sailors in the 24th Marine Expeditionary Unit participated in the evacuation effort.	• Various amphibious ships (USS *Iwo Jima*, USS *Whidbey Island*, USS *Trenton*, and USS *Nashville*) • Landing craft utility boats • Marine Corps CH-53 Super Stallion helicopters
South Sudan (2013)	On December 13, 2013, U.S. President Barack Obama ordered 45 combat-equipped U.S. military personnel to Juba, South Sudan, to protect U.S. citizens and property. DoD evacuated 120 personnel by air from the U.S. embassy in Juba to Nairobi, Kenya. Three days later, an additional 46 U.S. military personnel were deployed to Bor, South Sudan, to evacuate U.S. citizens and personnel residing there, but the operation was halted when U.S. aircraft came under fire.	• Two U.S. C-130 aircraft • Three CV-22 Ospreys
Libya (2014)	The U.S. embassy in Libya was evacuated in July 2014 following increased rebel violence. About eight diplomats and 200 security personnel were evacuated from Tripoli to Tunisia by ground convoy. The Air Force and Marine Corps provided air support for the convoy as a precaution.	• Ground convoy • Three F-17 aircraft • One CV-22 Osprey
Hurricane Irma (St. Maarten, 2017)	When Hurricane Irma hit St. Maarten, devastation followed, as well as civil unrest and crime. Three National Guard units from New York, Puerto Rico, and Kentucky evacuated 1,200 Americans from St. Maarten, flying them to safety in Puerto Rico over two days.	• Six C-130 aircraft • Two HH-60 helicopters
Sudan (2023)	The evacuations in Sudan have been multifaceted, complicated, and somewhat disorganized. The United States airlifted more than 100 people with Chinook helicopters on April 23, 2023. By April 30, 2023, DoS claimed that 1,000 Americans had been evacuated either by U.S. forces or U.S. allies and partners. The USNS *Brunswick* evacuated at least 300 people.	• Three Chinook Helicopters • USNS *Brunswick* (a fast transport vessel) • USS *Truxtin* (missile destroyer) • USS *Puller* (expeditionary vessel)

SOURCES: Schonauer, 2003; McEntee, 2003; Salazar Torreon and Plagakis, 2023; Schogol, 2006; "11,700 Americans Evacuated," 2006; Ford, 2007a; Markey, 2014; Martinez, 2017; Greer, 2007; Ware, 2023; Foggo, 2023; Gozzi and Davies, 2023; Ibrahim and Hatuqa, 2023; Shelbourne, 2023.

APPENDIX

Endnotes

1 Joint Publication 3-20, *Security Cooperation*, Joint Chiefs of Staff, May 23, 2017, p. GL-5.

2 Joint Publication 3-20, 2017, p. v.

3 Liana W. Rosen, "Security Cooperation Issues: FY2017 NDAA Outcomes," *In Focus*, Congressional Research Service, IF10582, version 2, January 6, 2017.

4 OSD, 2021; OSD, *Fiscal Year (FY) 2021 President's Budget: Justification for Security Cooperation Program and Activity Funding*, revision 1, April 2020; OSD, *Fiscal Year (FY) 2020 President's Budget: Justification for Security Cooperation Program and Activity Funding*, March 2019; OSD, *Fiscal Year (FY) 2019 President's Budget: Security Cooperation Consolidated Budget Display*, February 2018. DoD's budget justifications for security cooperation break down security cooperation funds by GCC but not by country. We considered using data from USAID's *U.S. Overseas Loans and Grants* publication, informally known as the Greenbook, which contains data on USG foreign assistance from 1945 to 2019 broken out by country. However, in comparing the funding accounts included in the Greenbook against the budget justifications, we found that the Greenbook does not include many of the funds DoD associates with security cooperation, such as funds for the Institute for Security Governance, DoD's primary mechanism for developing partner country institutional capabilities to govern, manage, operate, maintain, and sustain defense and security capabilities.

5 Joe Gould, "Eyeing China, Biden Defense Budget Boosts Research and Cuts Procurement," *Defense News*, May 28, 2021.

6 Bergen, Sterman, and Salyk-Virk, 2021; Serle and Purkiss, 2017.

7 New America, which draws on news reports, nongovernmental organization reports, and U.S. military press releases, presents civilian casualties as a range on the basis of the lowest and highest number of deaths reported across various sources. We used New America's high estimate of civilian casualties, which could result in an overestimation of condolence and solatia payments. However, we also note that these payments ultimately account for a small fraction of total estimated costs for targeted drone campaigns. As described by McNerney et al., there are differences in the sources and methods used by media outlets, nongovernmental organizations, and the U.S. military to identify and account for civilian harm resulting from U.S. operations. Therefore, estimates of civilian casualties for a given operation might differ significantly across sources, with U.S. confirmed casualties typically representing the lower end of this range (Michael J. McNerney, Gabrielle Tarini, Karen M. Sudkamp, Larry Lewis, Michelle Grisé, and Pauline Moore, *U.S. Department of Defense Civilian Casualty Policies and Procedures: An Independent Assessment*, RAND Corporation, RR-A418-1, 2021).

ABBREVIATIONS

AFRICOM	Africa Command
AOR	area of responsibility
CAP	combat air patrol
CENTCOM	Central Command
COIN	counterinsurgency
CONPLAN	contingency plan
COP	Conference of the Parties
DART	disaster assistance response team
DoD	Department of Defense
DoS	Department of State
EUCOM	European Command
FY	fiscal year
GCC	geographic combatant command
HADR	humanitarian assistance and disaster response
HQ	headquarters
INDOPACOM	Indo-Pacific Command
ISCP	International Security Cooperation Programs
MCO	major combat operation
NATO	North Atlantic Treaty Organization
NEO	noncombatant evacuation operation
NORTHCOM	Northern Command
O&M	operations and maintenance
OAIs	operations, activities, and investments
OHDACA	Overseas Humanitarian, Disaster, and Civic Aid
OPLAN	operational plan
OSD	Office of the Secretary of Defense

ROM	rough order of magnitude
RUMID	RAND U.S. Military Intervention Dataset
SSCI	significant security cooperation initiative
SOUTHCOM	Southern Command
SPP	State Partnership Program
UAE	United Arab Emirates
UN	United Nations
USAID	U.S. Agency for International Development
USG	U.S. government
USNS	U.S. Naval Ship
USS	U.S. Ship
VNSA	violent nonstate actor

REFERENCES

"11,700 Americans Evacuated from Lebanon," NBC News, July 24, 2006.

Al Fahaam, Tariq, and Hazem Hussein, "COP28 Will Be the UAE's Most Important Event in 2023: Sheikh Mohammed bin Rashid," Emirates News Agency, November 23, 2022.

Anderson, Cory T., Dave Blair, Mike Byrnes, Joe Chapa, Amanda Collazzo, Scott Cuomo, Olivia Garard, Ariel M. Schuetz, and Scott Vanoort, "Trust, Troops, and Reapers: Getting 'Drone' Research Right," War on the Rocks, April 3, 2018.

Barnett, Jon, and W. Neil Adger, "Climate Change, Human Security and Violent Conflict," *Political Geography*, Vol. 26, No. 6, August 2007.

Bergen, Peter, David Sterman, and Melissa Salyk-Virk, "America's Counterterrorism Wars: Tracking the United States's Drone Strikes and Other Operations in Pakistan, Yemen, Somalia, and Libya," New America, June 17, 2021. As of May 2023: https://www.newamerica.org/future-security/reports/americas-counterterrorism-wars/

Bureau of European and Eurasian Affairs, "NATO's Role in Bosnia and Herzegovina," fact sheet, U.S. Department of State, December 6, 2004.

Chandler, Nathan, Jeffrey Martini, Karen M. Sudkamp, Maggie Habib, Benjamin J. Sacks, and Zohan Hasan Tariq, *Pathways from Climate Change to Conflict in U.S. Central Command*, RAND Corporation, RR-A2338-2, 2023.

Christoff, Joseph, *Peacekeeping: Cost Comparison of Actual UN and Hypothetical U.S. Operations in Haiti*, Government Accountability Office, GAO-06-331, February 2006.

Cobble, W. Eugene, H. H. Gaffney, and Dmitry Gorenburg, *For the Record: All U.S. Forces' Responses to Situations, 1970–2000 (with Additions Covering 2000–2003)*, CNA Center for Strategic Studies, May 2005.

Department of Defense Instruction 5132.14, *Assessment, Monitoring, and Evaluation Policy for the Security Cooperation Enterprise*, Office of the Under Secretary of Defense for Policy, January 13, 2017.

Department of the Army, Office of the Assistant Secretary of the Army for Installations, Energy and Environment, *United States Army Climate Strategy*, February 2022.

DoD Open Government, "Security Cooperation," webpage, U.S. Department of Defense, undated. As of June 20, 2023: https://open.defense.gov/Transparency/Security-Cooperation/

Ellison, Tom, and Erin Sikorsky, "CCS Releases New Tool: Military Responses to Climate Hazards Tracker," Council on Strategic Risks, June 6, 2023.

Embassy of the United Arab Emirates, Washington, D.C., "A Shared Commitment to Climate Action," webpage, undated-a. As of May 3, 2023: https://www.uae-embassy.org/uae-us-cooperation/climate-energy

Embassy of the United Arab Emirates, Washington, D.C., "COP28," webpage, undated-b. As of June 19, 2023: https://www.uae-embassy.org/discover-uae/climate-energy/cop28

Farand, Chloé, "UAE Plans to Have It Both Ways as COP28 Climate Summit Host," Climate Home News, June 12, 2022.

Foggo, James G., "Evacuating Sudan: An Amphibious Gap and Missed Opportunity," *Defense News*, May 3, 2023.

Ford, Jess T., *State Department: The July 2006 Evacuation of American Citizens from Lebanon*, Government Accountability Office, GAO-07-893R, June 7, 2007a.

Ford, Jess T., *State Department: Evacuation Planning and Preparations for Overseas Posts Can Be Improved*, Government Accountability Office, GAO-08-23, October 2007b.

Frederick, Bryan, Jennifer Kavanagh, Stephanie Pezard, Alexandra Stark, Nathan Chandler, James Hoobler, and Jooeun Kim, *Assessing Trade-Offs in U.S. Military Intervention Decisions: Whether, When, and with What Size Force to Intervene*, RAND Corporation, RR-4293-A, 2021. As of June 14, 2023: https://www.rand.org/pubs/research_reports/RR4293.html

Gould, Joe, "Eyeing China, Biden Defense Budget Boosts Research and Cuts Procurement," *Defense News*, May 28, 2021.

Government Communications Office, State of Qatar, "Environment and Sustainability," webpage, undated. As of May 4, 2023: https://www.gco.gov.qa/en/focus/environment-and-sustainability/

Gozzi, Laura, and Alys Davies, "Sudan Fighting: Diplomats and Foreign Nationals Evacuated," BBC News, April 24, 2023.

Greer, Dale, "Airlift Operation Evacuates 1,000 U.S. Citizens from St. Maarten; 17 Kentucky Air Guardsmen Returning Home Today, 45 More Deploying to Support Irma Rescue Missions," 123rd Airlift Wing Public Affairs, September 11, 2017.

Hasik, James, "Affordably Unmanned: A Cost Comparison of the MQ-9 to the F-16 and A-10, and a Response to Winslow Wheeler's Criticism of the Drone," *James Hasik: Thinking on Innovation, Industry, and International Security* blog, June 20, 2012. As of September 28, 2023: https://www.jameshasik.com/weblog/2012/06/affordably-unmanned-a-cost-comparison-of-the-mq-9-to-the-f-16-and-a-10-and-a-response-to-winslow-whe.html

Ibrahim, Arwa, and Dalia Hatuqa, "Sudan Updates: Warring Sides Agree to Extending Truce," Al Jazeera, April 30, 2023.

Jacobsen, Rowan, "Israel Proves the Desalination Era Is Here," *Scientific American*, July 29, 2016.

Joint and Coalition Operational Analysis, "Operation UNITED ASSISTANCE: The DoD Response to Ebola in West Africa," January 6, 2016.

Joint Publication 3-20, *Security Cooperation*, Joint Chiefs of Staff, May 23, 2017.

Joint Publication 3-29, *Foreign Humanitarian Assistance*, Joint Chiefs of Staff, May 14, 2019.

Kavanagh, Jennifer, Bryan Frederick, Matthew Povlock, Stacie L. Pettyjohn, Angela O'Mahony, Stephen Watts, Nathan Chandler, John Speed Meyers, and Eugeniu Han, *The Past, Present, and Future of U.S. Ground Interventions: Identifying Trends, Characteristics, and Signposts*, RAND Corporation, RR-1831-A, 2017. As of June 14, 2023: https://www.rand.org/pubs/research_reports/RR1831.html

Kavanagh, Jennifer, Bryan Frederick, Alexandra Stark, Nathan Chandler, Meagan L. Smith, Matthew Povlock, Lynn E. Davis, and Edward Geist, *Characterizations of Successful U.S. Military Interventions*, RAND Corporation, RR-3062-A, 2019. As of July 13, 2023: https://www.rand.org/pubs/research_reports/RR3062.html

Kim, Julie, *Bosnia Implementation Force (IFOR) and Stabilization Force (SFOR): Activities of the 104th Congress*, Congressional Research Service, No. 96-723, January 6, 1997.

Kingdon, Ashton, and Briony Gray, "The Class Conflict Rises When You Turn Up the Heat: An Interdisciplinary Examination of the Relationship Between Climate Change and Left-Wing Terrorist Recruitment," *Terrorism and Political Violence*, Vol. 34, No. 5, 2022.

Lakhani, Nina, "'We Couldn't Fail Them': How Pakistan's Floods Spurred Fight at COP for Loss and Damage Fund," *The Guardian*, November 20, 2022.

Markey, Patrick, "U.S. Evacuates Libya Embassy After 'Free-Wheeling Militia Violence,'" Reuters, July 27, 2014.

Martinez, Luis, "Americans on St. Maarten Tell of Irma's Devastation, Lawlessness; 1,200 Evacuated," ABC7 News, September 10, 2017.

Mavrakou, Stefanie, Emelie Chace-Donahue, Robin Oluanaigh, and Meghan Conroy, "The Climate Change–Terrorism Nexus: A Critical Literature Review," *Terrorism and Political Violence*, Vol. 34, No. 5, 2022.

McAndrew, Anne J., "Fiscal Year (FY) 2019 Department of Defense (DoD) Fixed Wing and Helicopter Reimbursement Rates," memorandum to assistant secretaries, deputy chiefs, directors of the military departments, et al., Office of the Under Secretary of Defense (Comptroller), October 12, 2018.

McEntee, Marni, "USAFE Sends Airmen to Aid Liberia Mission," *Stars and Stripes*, July 15, 2003.

McNerney, Michael J., Gabrielle Tarini, Karen M. Sudkamp, Larry Lewis, Michelle Grisé, and Pauline Moore, *U.S. Department of Defense Civilian Casualty Policies and Procedures: An Independent Assessment*, RAND Corporation, RR-A418-1, 2021. As of June 19, 2023: https://www.rand.org/pubs/research_reports/RRA418-1.html

Melito, Thomas, *UN Peacekeeping: Cost Estimate for Hypothetical U.S. Operation Exceeds Actual Costs for Comparable UN Operation*, Government Accountability Office, GAO-18-243, February 2018.

Ministry of Defense, Netherlands, "Kosovo Force 1999–2000 (KFOR)," webpage, undated. As of May 2023: https://english.defensie.nl/topics/historical-missions/mission-overview/1999/kosovo-force-1999-%E2%80%93-2000-kfor

Ministry of Foreign Affairs, Government of Pakistan, "Pakistan Welcomes the Historic Decision of COP27 to Establish the Fund for Loss and Damage," press release, November 20, 2022.

Miro, Michelle E., Flannery Dolan, Karen M. Sudkamp, Jeffrey Martini, Karishma V. Patel, and Carlos Calvo Hernandez, *A Hotter and Drier Future Ahead: An Assessment of Climate Change in U.S. Central Command*, RAND Corporation, RR-A2338-1, 2023.

Mohseni-Cheraghlou, Amin, "Fossil Fuel Subsidies and Renewable Energies in MENA: An Oxymoron?" Middle East Institute, February 23, 2021.

Moroney, Jennifer D. P., Stephanie Pezard, Laurel E. Miller, Jeffrey Engstrom, and Abby Doll, *Lessons from Department of Defense Disaster Relief Efforts in the Asia-Pacific Region*, RAND Corporation, RR-146-OSD, 2013. As of March 22, 2023: https://www.rand.org/pubs/research_reports/RR146.html

National Guard Bureau, "National Guard State Partnership Program," map, May 17, 2023.

NATO—*See* North Atlantic Treaty Organization.

North Atlantic Treaty Organization, "NATO Mission in Kosovo (KFOR)," webpage, undated. As of May 2023: https://shape.nato.int/ongoingoperations/nato-mission-in-kosovo-kfor-

NTS-Asia, "About Non-Traditional Security," webpage, undated. As of August 10, 2023: https://rsis-ntsasia.org/about-nts-asia/

Office of the Secretary of Defense, *Fiscal Year (FY) 2019 President's Budget: Security Cooperation Consolidated Budget Display*, February 2018.

Office of the Secretary of Defense, *Fiscal Year (FY) 2020 President's Budget: Justification for Security Cooperation Program and Activity Funding*, March 2019.

Office of the Secretary of Defense, *Fiscal Year (FY) 2021 President's Budget: Justification for Security Cooperation Program and Activity Funding*, revision 1, April 2020.

Office of the Secretary of Defense, *Fiscal Year (FY) 2022 President's Budget: Justification for Security Cooperation Program and Activity Funding*, May 2021.

Office of the Under Secretary of Defense (Acquisition and Sustainment), *Department of Defense Climate Adaptation Plan*, U.S. Department of Defense, September 1, 2021.

OSD—*See* Office of the Secretary of Defense.

Palik, Júlia, Siri Aas Rustad, Kristian Berg Harpviken, and Fredrik Methi, *Conflict Trends in the Middle East, 1989–2019*, Peace Research Institute Oslo, 2020.

Peterson, Heather, "U.N. Peacekeeping Is a Good Deal for the U.S.," *RAND Blog*, April 2, 2017. As of September 28, 2023: https://www.rand.org/blog/2017/04/un-peacekeeping-is-a-good-deal-for-the-us.html

Rosen, Liana W., "Security Cooperation Issues: FY2017 NDAA Outcomes," *In Focus*, Congressional Research Service, IF10582, version 2, January 6, 2017.

Salazar Torreon, Barbara, and Sofia Plagakis, *Instances of Use of United States Armed Forces Abroad, 1798–2023*, Congressional Research Service, R42738, version 39, March 8, 2022, updated June 7, 2023.

Saudi & Middle East Green Initiatives, "SGI Target: Reduce Carbon Emissions by 278 mtpa by 2030," webpage, undated. As of May 4, 2023: https://www.greeninitiatives.gov.sa/about-sgi/sgi-targets/reducing-emissions/reduce-carbon-emissions

Schogol, Jeff, "About 1,200 Americans Evacuated from Lebanon," *Stars and Stripes*, July 20, 2006.

Schonauer, Scott, "Troops Boost Security at African Hot Spots," *Stars and Stripes*, June 14, 2003.

Serle, Jack, and Jessica Purkiss, "Drone Wars: The Full Data," database, Bureau of Investigative Journalism, January 1, 2017. As of May 2023: https://www.thebureauinvestigates.com/stories/2017-01-01/drone-wars-the-full-data

Shatz, Howard J., Karen M. Sudkamp, Jeffrey Martini, Mohammad Ahmadi, Derek Grossman, and Kotryna Jukneviciute, *Mischief, Malevolence, or Indifference? How Competitors and Adversaries Could Exploit Climate-Related Conflict in the U.S. Central Command Area of Responsibility*, RAND Corporation, RR-A2338-4, 2023.

Shelbourne, Mallory, "U.S. Navy Sends Nontraditional Ships to Support Sudan Evacuation," USNI News, May 1, 2023.

Sim, Li-Chen, "Renewable Power Policies in the Arab Gulf States," Middle East Institute, February 8, 2022.

Sletzinger, Martin, "Iraq Through the Lens of Bosnia and Kosovo," Wilson Center, March 17, 2003.

Surkes, Sue, "Israel, Jordan, UAE Sign New MOU on Deal to Swap Solar Energy for Desalinated Water," *Times of Israel*, November 8, 2022.

Toukan, Mark, Stephen Watts, Emily Allendorf, Jeffrey Martini, Karen M. Sudkamp, Nathan Chandler, and Maggie Habib, *Conflict Projections in U.S. Central Command: Incorporating Climate Change*, RAND Corporation, RR-A2338-3, 2023.

Turse, Nick, Henrik Moltke, and Alice Speri, "Secret War," The Intercept, June 20, 2018.

UN—*See* United Nations.

United Arab Emirates Government, "UAE Energy Strategy 2050," webpage, updated August 14, 2023. As of September 28, 2023: https://u.ae/en/about-the-uae/strategies-initiatives-and-awards/strategies-plans-and-visions/environment-and-energy/uae-energy-strategy-2050

United Nations Integrated Mission in Timor-Leste, "UNMIT Facts and Figures," fact sheet, undated.

United Nations Mission in Côte d'Ivoire, "Côte d'Ivoire—MINUCI—Facts and Figures," fact sheet, United Nations Peacekeeping, 2004.

United Nations Mission in Ethiopia and Eritrea, "Ethiopia and Eritrea—UNMEE—Facts and Figures," fact sheet, United Nations Peacekeeping, 2009.

United Nations Mission in Sierra Leone, "Sierra Leone—UNAMSIL—Facts and Figures," fact sheet, 2005.

United Nations Mission in the Central African Republic and Chad, "MINURCAT Facts and Figures," fact sheet, United Nations Peacekeeping, undated.

United Nations Mission in the Sudan, "UNMIS Facts and Figures," fact sheet, undated.

United Nations Mission of Support in East Timor, "East Timor—UNMISET—Facts and Figures," fact sheet, United Nations Peacekeeping, 2005.

United Nations Operation in Burundi, "Burundi—ONUB—Facts and Figures," fact sheet, United Nations Peacekeeping, 2007.

United Nations Organization Mission in the Democratic Republic of the Congo, "MONUC Facts and Figures," fact sheet, United Nations Peacekeeping, undated.

United Nations Peacekeeping, "Promoting Human Rights," webpage, undated. As of August 10, 2023: https://peacekeeping.un.org/en/promoting-human-rights

United Nations Supervision Mission in Syria, "UNSMIS Facts and Figures," fact sheet, undated.

United Nations Transitional Administration in East Timor, "East Timor—UNTAET: Facts and Figures," fact sheet, United Nations Peacekeeping, 2002.

U.S. Agency for International Development, "Kosovo Crisis Fact Sheet #22," fact sheet, April 11, 1999a.

U.S. Agency for International Development, "Kosovo Crisis Fact Sheet #37," fact sheet, April 26, 1999b.

U.S. Agency for International Development, "Kosovo Crisis Fact Sheet #144," fact sheet, April 7, 2000a.

U.S. Agency for International Development, "East Timor—Crisis Fact Sheet Summary #1, Fiscal Year (FY) 2000," fact sheet, October 3, 2000b.

U.S. Agency for International Development, "Afghanistan Complex Emergency Situation Report #04 (FY 2003)," March 3, 2003.

U.S. Agency for International Development, "Lebanon: Complex Emergency Information Bulletin # 1 (FY 2006)," July 21, 2006.

U.S. Agency for International Development, "Sri Lanka—Complex Emergency Fact Sheet #1, Fiscal Year (FY) 2007," fact sheet, February 2, 2007a.

U.S. Agency for International Development, "Lebanon Humanitarian Emergency: USG Humanitarian Situation Report # 11 (FY) 2007," June 20, 2007b.

U.S. Agency for International Development, "Global—Influenza A/H1N1 Fact Sheet #1, Fiscal Year (FY) 2009," fact sheet, May 5, 2009a.

U.S. Agency for International Development, "Global—Influenza A/H1N1 Fact Sheet #3, Fiscal Year (FY) 2009," fact sheet, May 18, 2009b.

U.S. Agency for International Development, "Georgia: Complex Emergency Fact Sheet #2, Fiscal Year (FY) 2009," fact sheet, June 18, 2009c.

U.S. Agency for International Development, "Pakistan—Complex Emergency Fact Sheet #31, Fiscal Year (FY) 2009," fact sheet, September 22, 2009d.

U.S. Agency for International Development, "Sri Lanka—Complex Emergency Fact Sheet #18, Fiscal Year (FY) 2009," fact sheet, September 30, 2009e.

U.S. Agency for International Development, "Iraq—Complex Emergency Fact Sheet #1, Fiscal Year (FY) 2012," fact sheet, October 21, 2011.

U.S. Agency for International Development, "Iraq—Complex Emergency Fact Sheet #2, Fiscal Year (FY) 2020," fact sheet, May 8, 2020.

U.S. Agency for International Development, "Central Asia Region Complex Emergency Situation Report #41 (FY 2002)," August 16, 2022.

U.S. Agency for International Development, "Pakistan—Floods Factsheet #2, Fiscal Year (FY) 2023," fact sheet, January 12, 2023.

USAID—*See* U.S. Agency for International Development.

U.S. Central Command, "Operations and Exercises," webpage, undated. As of July 13, 2023: https://www.centcom.mil/OPERATIONS-AND-EXERCISES/

U.S. Department of Defense, *Summary of the 2018 National Defense Strategy of the United States of America*, 2018.

von Lossow, Tobias, *Water as Weapon: IS on the Euphrates and Tigris*, German Institute for International and Security Affairs, January 2016.

Ware, Doug G., "Navy Moves Ships to Red Sea in Case US Needs to Move Americans Out of Sudan," *Stars and Stripes*, April 24, 2023.

White House, "Winning the War and the Peace in Kosovo," undated.

Zenko, Micah, "What Does Libya Cost the United States?" *Politics, Power, and Preventive Action*, Council on Foreign Relations blog, August 11, 2011. As of September 28, 2023: https://www.cfr.org/blog/what-does-libya-cost-united-states

Zyla, Benjamin, *Sharing the Burden? NATO and Its Second-Tier Powers*, University of Toronto Press, 2015.

PHOTO CREDITS: Cover: Pakistan humanitarian aid during flood, Sgt. Jason Bushong/U.S. Army; page ix: Pakistan flooding, UK Department for International Development (CC BY 2.0, via Wikimedia Commons); page 28: UK Department for International Development (CC BY 2.0, via Wikimedia Commons).